D1211840

Hidden Connections
and Double Meanings

A MATHEMATICAL EXPLORATION

David Wells

Dover Publications, Inc.
Mineola, New York

This book is dedicated to Guinevere

Bibliographical Note

This Dover edition, first published in 2018, is an unabridged republication of *Hidden Connections, Double Meanings*, originally published in 1988 by the Cambridge University Press, Cambridge.

Library of Congress Cataloging-in-Publication Data

Names: Wells, D. G. (David G.), author. | Wells, D. G. (David G.). Hidden connections, double meanings.
Title: Hidden connections and double meanings : a mathematical exploration / David Wells.
Other titles: Hidden connections, double meanings
Description: Dover edition. | Mineola, New York : Dover Publications, Inc., 2018. | Originally published: Cambridge : Cambridge University Press, 1988.
Identifiers: LCCN 2018012325 | ISBN 9780486824628 | ISBN 0486824624
Subjects: LCSH: Problem solving. | Mathematical recreations.
Classification: LCC QA63 .W4 2018 | DDC 793.7—4dc23
LC record available at https://lccn.loc.gov/2018012325

Manufactured in the United States by LSC Communications
82462401   2018
www.doverpublications.com

Contents

Mathematics, riddles and humour

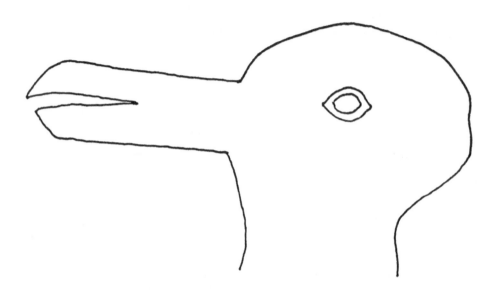

This idea was invented by the psychologist, Jastrow. Is it a duck, or is it a rabbit? Or is it neither – merely some lines on a flat sheet of paper which our brains interpret as one or the other?

Art is full of such tricks. Indeed, all traditional art depends on such illusions. We 'see' a person in the flat paint of their portrait. The painter lays brush strokes of paint onto the flat surface of the canvas and we see a bowl of fruit, or skaters in a landscape. How clever we are! Indeed, our brains are marvellously adept at seeing surprising relationships between images, or seeing the same image in different ways.

Sometimes our own brains deceive us and lead us astray. We make fools of ourselves by being too clever. Yet without our natural talent we would never be able to make sense of the world.

How else would we 'see' that these four women, although clearly of different sizes on the surface of the paper, are 'really' the same size?

distorted

distorted

distorted

distorted

DISTORTED

distorted

distorted

DISTORTED

How can we know that all these squiggles say the same word 'distorted'?

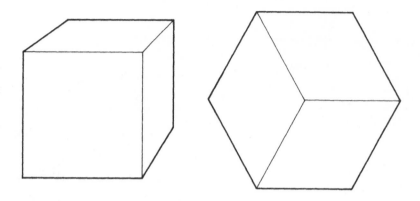

The result of looking often depends on **how** you look. Only from a special angle will this cube look like a hexagon.

8. The largest of the Rocs picks her up by the skirt.

Can you see the upside-down figure in the drawing by Gustave Verbeek, without turning the page around? Not so easy! Although there are two distinct images on the paper, only one of them is 'staring you in the face'.

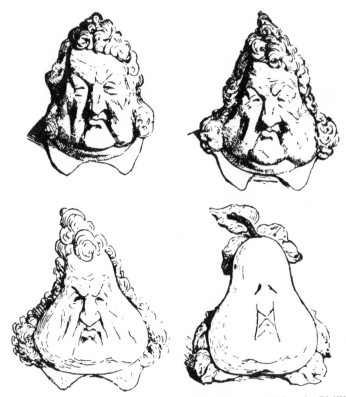

The nineteenth-century French cartoonist who turned Louis Phillippe's fleshy features into a pear also expressed a second, double meaning; 'poire' was slang for a simpleton.

So much humour is formed on a skeleton of double meanings, mistaken meanings, incongruity, surprise, and the familiar in strange dress. Even the simplest riddles incorporate these features:

'What is small and green, and zooms down the M1 at 90 miles per hour?'

When I was a schoolboy, the answer was an Austin Healey Sprout. This car is now an antique, but childish ingenuity is not defeated. The contemporary answer is a lettuce Elan!

The ambiguity of common words is exploited again and again:

'What has four wheels and flies?' 'A dust cart!'

'Time flies!' 'You cannot – they move too fast!'

Wit and humour, once discovered, accompany us throughout our lives. I still remember the very first joke I heard, or rather, the first joke that I understood. I recall feeling highly pleased with myself. I had made an important discovery. It went like this: 'Woman to assistant, "There's a ladder in this stocking!" Assistant, "What do you expect for that price? A marble staircase?"' Children today laugh at: 'What do you call a camel with three humps?' 'Humphrey!'

Adults claim not to enjoy such trifles, though during the 1976 drought

7

everyone laughed at 'Save water – bath with a friend' and even adults are allowed to smile at snappy graffiti.

Sophisticated adults' dislike of riddles, if indeed it is real and not feigned, is a recent phenomenon. From Homer, who is supposed to have died of frustration on failing to solve 'What we caught we threw away, what we could not catch, we kept,' (fleas!) to our Victorian great great grandparents, riddling and word play were respectable pastimes for politicians as well as poets.

Perhaps adults are pretending. Perhaps they are foolishly ashamed to share anything with children. There are certainly plenty of word plays and double meanings in the sole feature shared by the *Guardian* and *Beano,* the strip cartoon.

This one has a mathematical flavour.

Mathematics has much in common with riddling, and with humour. Everything in mathematics has many meanings. Every diagram and every figure, every sum and every equation, can be 'seen' in different ways. Every sentence, in English or in algebra, can be variously read and interpreted.

This is one reason why mathematics is so powerful. The mathematician who spots that two apparently different problems are 'the same', solves both by solving one. Realise that a million problems are 'essentially' the same, and you can solve a million problems by solving one. Now there is power indeed!

Mathematics, riddles and humour have something else in common. They share similar emotions. Humour, of course, is quick. No one laughs at a joke which has taken an hour to work out, and a joke that has to be explained is an embarrassment to the comedian and the audience. Riddles are harder work. Mathematics – and science – are harder work still, but even more enjoyable.

Watch someone solve a problem. Look at their face as the penny drops! They will very probably smile, or even laugh out loud! When I was a child and I solved a tricky problem, I could not sit still, but got up and walked excitedly round the room, or went to get something to eat. I still feel the same way.

This book is about solving mathematical riddles by 'seeing', sometimes with the eyes, sometimes without. 'Seeing' is a metaphor, which does not always have to be taken literally. We can see with our brains, with our minds. In mysterious ways which we hardly start to understand, our brains work below our conscious thoughts, making sense, making connections, and frequently surprising us.

Here is a first chance to let the penny drop gently with a classic problem.

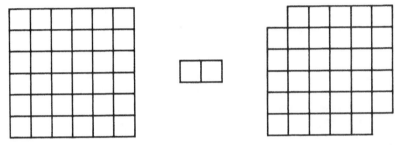

Can the square on the left be filled with 18 of the small dominoes? Yes, easily, by filling each row with three dominoes in a line. What about the right-hand square which has two corners missing? Can this be filled with one fewer dominoes? Try it and decide!

The secret of this puzzle is to think of the square as a chequerboard, with two white squares missing from the opposite corners. Now we can see that *any* domino will cover one black square and one white square no matter how or where we place it.

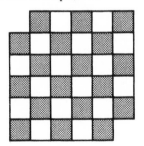 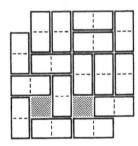

We can therefore be absolutely confident that the board with the missing corners can never be covered, however many attempts we make. The diagram on the right is merely to confirm for those of little faith that the extra two black squares do appear in this instance as the unfillable squares!

Problems

1

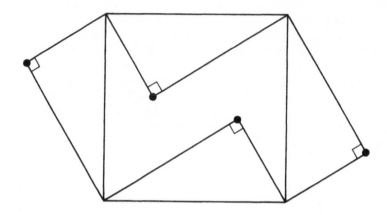

Four identical right-angled triangles have been added to a square, two on the outside, two on the inside. A straight-edge will confirm that the four marked points lie on a straight line – but why?

2 What is the largest number of faces of a cube that you could possibly see at one time, if the faces are not transparent?

3

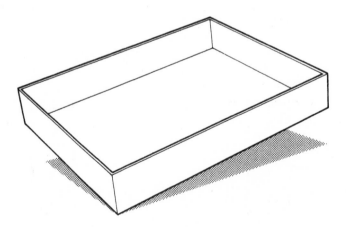

This rectangular box has one corner resting on the table top. What is the connection between the heights of the other three corners of its base above the table?

4 This is an unusual way to subtract 153 from 482: 482 282

153   47
───────
329

Why does this method work?

5

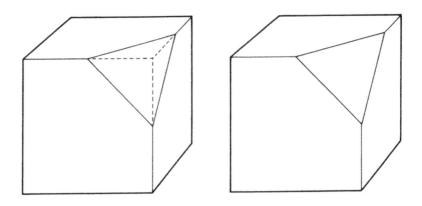

Slice a cube with a flat knife stroke and a new face will be created. These two figures show one way to get an equilateral triangle.

Which of the following shapes could also be created by slicing a cube? (A few experiments with a small cube of cheese may be of assistance!)

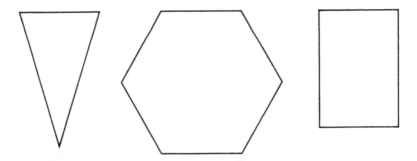

6 If the following knot is tied in a long thin strip of paper with parallel sides and then carefully flattened, what regular polygon will appear?

Exploding the image

Here is a key and a lock. The key turns in the lock and the lock opens. How?
That is the puzzle! If a diagram of the lock were pinned to the wall beside it,
or if the lock were made of glass, it would be only too easy to pick it or to
make a duplicate key. The lock keeps its secret to the uninitiated because its
mechanism is hidden inside.

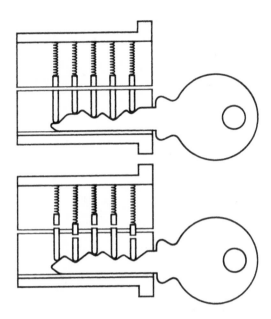

1.2 The lock contains several short pins each containing a small gap. When the correct key is inserted, these gaps are aligned and the cylinder will turn, withdrawing the bolt. The wrong key fails to align the pins.

Now that the lock has been exposed to the light of day, the mechanism is quite easy to follow. The connections are easy to make. Moreover, having seen how one lock mechanism works, it should be easier to understand other lock mechanisms, at least those of a similar type.

(What do we mean when we say that two locks are of the same type? That is a problem in itself!)

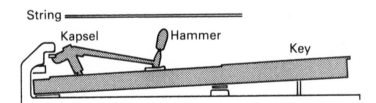

1.3 As the player presses the pivoted key, the other end rises like a see-saw. One end of the pivoted hammer bar is held down, forcing the hammer up to hit the string.

The ingenious mechanism of a piano allows the player to strike many notes rapidly in succession, strongly or lightly, damped or undamped. The levers which compose the mechanism are not peculiar to pianos, but are found everywhere, including the lock already described.

13

Geometrical problems have traditionally demanded that the solver makes a connection. These connections are often hidden and obscure, and there are no laws or rules which tell the solver what to do. This geometrical problem requires the solver to make a connection, literally by drawing in some missing lines, and metaphorically by spotting a relationship.

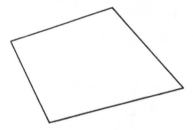

This diagram is intended to represent a general quadrilateral, that is, any quadrilateral at all. The problem is to explain why the four interior angles of this quadrilateral add up to a total of 360°. Any doubting readers with a protractor at hand can check that they do so. To the nearest degree, the angles are 81°, 115°, 62° and 102°, which total to 360°.

Experiments with several quadrilaterals will show that their angles undoubtedly do sum to 360°, give or take a fraction of a degree, and no mathematician worth his or her salt will easily doubt that the difference can be put down to 'experimental error'. The question is, 'Why?'

Since all arguments and conclusions require premises, we shall take one clue as a starting point. This clue is that the three angles of any triangle sum to 180°.

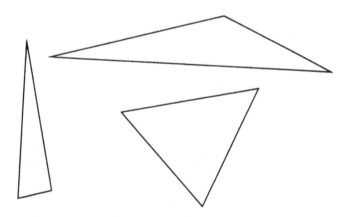

The word 'clue' is appropriate. A clue, or clew, was originally a ball of thread. The Greek hero Theseus who ventured into King Minos' labyrinth to seek the Minotaur and kill it, escaped from the maze because he had unwound a ball of thread on his way in, and by this clue he found his way out again.

How can the clue about the angles of any triangle be exploited?

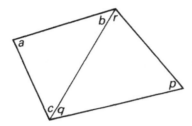

The original quadrilateral can now be seen as two triangles. Their six angles, *a*, *b*, *c* and *p*, *q*, *r*, together make up the original four angles of the quadrilateral, *a*, *b+r*, *p* and *c+q*, although two of these have been split into two pieces each. It is no surprise that if the angles of one triangle sum to 180°, the angles of two triangles will sum to twice as much.

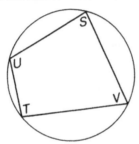

The idea of dividing a quadrilateral into triangles can be used as a lever to discover more properties of quadrilaterals. This diagram shows a rather special quadrilateral, special because it is inscribed in a circle. Because of this extra feature it has an extra property. It is still true that its angles sum to 360°, as near as practical measurement will allow, but now it is also true that pairs of opposite angles sum to 180°, $S + T = 180°$ and $U + V = 180°$.

Why is this? The reason must be something to do with the circle, because the original quadrilateral, which cannot be fitted into a circle, lacks this property. What difference might the circle make? What is so special about a circle, or the four corners of the quadrilateral which lie on it?

A circle is the path of a point which moves so that it is always the same distance from a fixed point, its centre. This extra clue can only be used by marking the centre of the circle and drawing the lines of equal length which join it to the four vertices. At once the diagram takes on a different complexion. As before, the original quadrilateral has been dissected into triangles, but this time there are four rather than two.

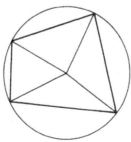

Where do we go from here? What will our next move be? Somehow we must exploit the fact that the radii of the circle, the lines from the centre to the vertices, are equal in length. Of course! The triangles are what are called isosceles, because they have a pair of sides of equal length. It follows that they have a pair of equal angles also, which are marked with matching symbols in this figure.

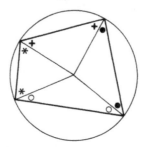

The circle, having been fully exploited, can now be forgotten. Looking carefully at the quadrilateral we see that the pair of angles marked S and T is composed of one each of the angles marked with the four symbols. Likewise, the angles marked U and V are also composed of one each of the marked angles, but in a different order. No wonder that $S + T = U + V$!

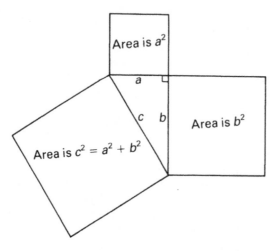

Area is a^2

a

c b

Area is b^2

Area is $c^2 = a^2 + b^2$

These properties of a quadrilateral are relatively easy to 'see'. Other geometrical properties are more subtle. The theorem of Pythagoras says that if squares are drawn on the sides of a right-angled triangle then the area of the largest square (the one on the 'hypotenuse'), is equal to the sum of the areas of the squares on the other two sides. It has been famous for more than 2000 years and more than 200 proofs of this remarkable theorem have been published, including one by President Garfield of the United States.

How can we prove it? A simple and plausible idea is to cut up the two small squares into several pieces which will physically fit together to make the large square. Any mathematician will naturally think first of performing this dissection symmetrically. For example, the small square could be placed in one piece at the centre of the large square, and the surrounding area dissected into pieces to make up the middle square.

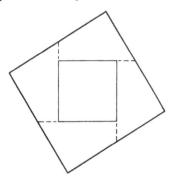

By imagining the small square set skew, with cuts perhaps following these dotted lines, this seems a reasonable ambition, because it is easy to cut up the middle square into pieces of roughly the right shape, like this.

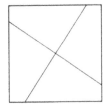

Is it possible to get exactly the right shapes? Yes, it is, if the length PQ is marked along the edge at R, and the cut starts at X, half way between R and the corner.

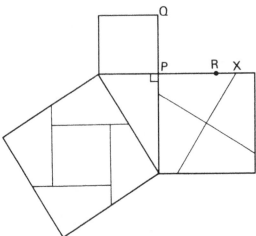

The five pieces now fit perfectly into the 'square on the hypotenuse'. Why does the cut at X always work? That is indeed a gaping hole in this solution, which will be filled in due course.

17

Problems

1

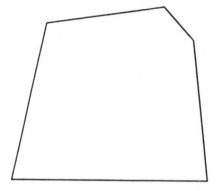

What is the sum of the angles of any convex pentagon, like this?

2

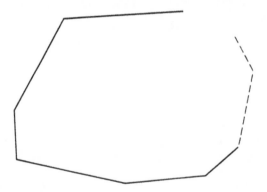

What is the rule for the sum of the angles of a convex polygon with n sides?

3

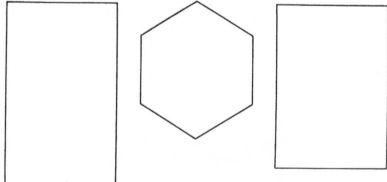

Which of these shadows could be the shadow of a cube, from a light sufficiently far away?

4

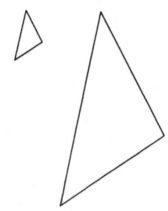

Where is the stack of cubes in this tesselation of hexagons?

5

These triangles are similar in shape, but the larger is 4 times as long in every direction as the smaller. How many of the smaller could be packed into the larger?

6

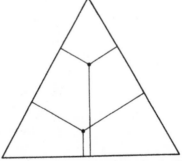

In this equilateral triangle a point has been marked and the lines which go directly from the point to the three sides, meeting them at right-angles. As experiment will confirm, the total length of these three lines does not depend on the position of the point originally chosen. Why not?

The hidden image

Where are the animals in this scene? Why are they so difficult to spot? Would you spot them at all, if you were not told that they were there? Why is there such a difference between what your eyes see, and what your brain sees?

Where is the five-pointed star in this jagged design? It lies near the bottom left-hand corner, but few people spot this immediately although it is the only 'regular' bit of the design.

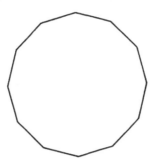

That design had no pattern. This symmetrical and elegant figure in contrast is a dodecagon, a regular 12-sided figure, also called a regular 12-gon.

It contains within itself, as it were, two regular hexagons, three squares and four equilateral triangles. Where?

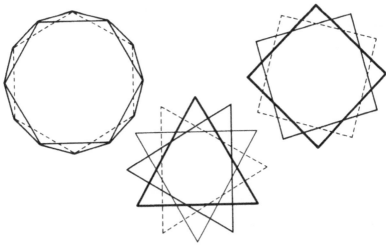

There is a neat relationship between the regular polygons which can be 'seen' in the dodecagon, and the factors of 12, the numbers which divide 12 without remainder.

21

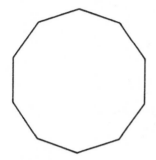
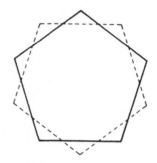

This drawing may look very similar at first sight to the dodecagon, but on counting its edges it has only 10. Only one regular polygon can be seen in the decagon; the regular pentagon appears twice.

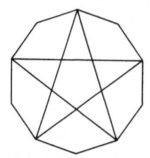

However, there are two regular star polygons. The 5-pointed star or pentagram is drawn by starting at one vertex and drawing a straight line to the fourth vertex on, and continuing until you get back to the vertex you started at. For the 10-pointed star you draw lines to the third vertex on. These star figures are harder to follow with the eye because they intersect themselves. They have a slightly three-dimensional quality.

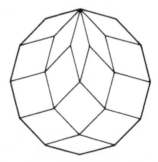

This dodecagon is quite flat, and has been dissected into diamonds.

The diamond tiles are very like the children's coloured wooden tiles that are sold in a rectangular tray and can be used to make a great variety of different designs. There are several other ways in which these 15 tiles can be fitted into the dodecagon. This is one of them:

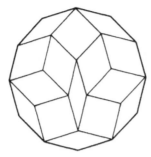

This dissection could also have been found by starting with the dodecagon, and filling in diamonds, starting from any pair of adjacent edges. It turns out that there are a number of choices on the way, points at which more than one diamond may be completed. Some of these choices allow the pattern to be completed, others lead to a dead end. Curiously, every completed pattern contains 6 of the narrowest diamonds, 6 of the wider diamonds, and 3 squares. No other combination is possible.

Either by calculation or measurement, it is apparent that there is a simple relationship between the diamonds of each size. The angles of the narrowest are 30-150-30-150. The middling size is 60-120-60-120 and the squares of course are 90-90-90-90.

This simplicity suggests very strongly to the mathematician that there is something 'natural' about this dissection, and it is no surprise that it can be used to dissect a dodecagon elegantly into other shapes.

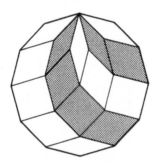

In particular, if the diamonds are thought of as strips, as shown, then four dodecagons can be fitted together to make one large dodecagon in several ways, as these exquisite patterns show.

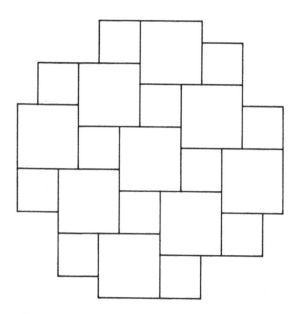

This floor tiling is well known. The squares may be of any two different sizes. At first sight, the tiling may seem to be composed of two different units, the squares of two sizes. Is there a way of seeing it as a tiling of identical units of one size only?

Yes, there is. Concentrate on the large squares as they go marching in sloping lines across the floor. Look at them marching in one direction, steeply down to the right. Then look at them marching more gently up to the top right.

The pattern stands out more strongly if they are shaded, or if you imagine that the small squares are simply spaces you can look through.

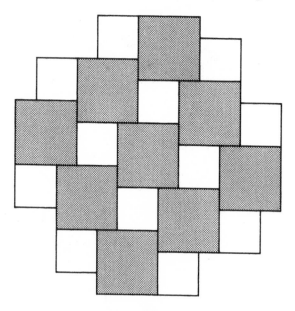

The pattern can also be emphasised by choosing 'the same' point in each large square, and focussing on these corresponding points.

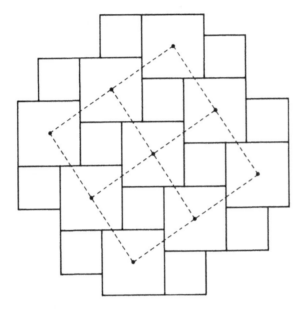

Here the centre of each large square has been chosen, and these points have been joined together, following the march of the squares. The result is indeed a pattern of even larger squares joined edge to edge in the usual manner.

We can now think of the pattern as composed of these very large squares repeated, each giant square having the same design drawn upon it, of one small square, and four identical pieces from the middle-sized square. Moreover, the bits of the middle sized square all fit together to make that square exactly, as this picture shows.

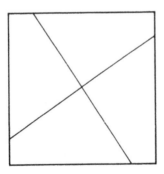

However, this picture is perhaps unnecessary, because the identical picture has already appeared in Chapter 1, where it was a means of proving the theorem of Pythagoras. This tiling fills in the gap in that proof, by explaining why and exactly where the cuts must be made in the middle-sized square.

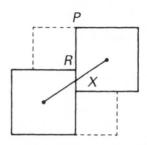

When the centres of two large squares are joined, the line naturally passes symmetrically through their common length of edge and bisects it at X. Since both angles are exactly filled by one of the smaller squares, the length from P to R is equal to the side of the smaller square.

Joining the centres of the large squares is natural, and naturally produces a symmetrical pattern. What would have happened if another set of corresponding points had been chosen?

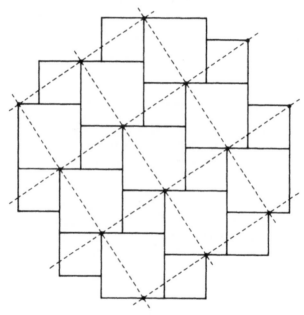

By choosing the top right corner of each large square and joining the points together, the pattern of giant squares still appears, but shifted upwards and to the right, as it were.

What about the dissection? Will the parts of the giant square still make up both the smaller squares? Certainly, and still in five pieces, though the small square is now divided into two and the large square into only three pieces.

The fact is that it makes no difference at all which set of corresponding points is chosen. The same pattern of giant squares appears, and each giant square dissects into at most nine pieces which will make the other two squares.

So by this method alone there are an infinite number of 'different' proofs of Pythagoras' theorem!

Problems

1

Make a tessellation of identical pentagons by placing one of these tessellations of elongated hexagons over another identical tessellation.

2

How can this Greek cross be dissected into several pieces which will fit together to form a square?

3

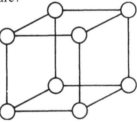

Colour the eight vertices of a cube with two colours, each used four times, so that no face can be distinguished from any other face by its colouring.

4

Imagine that in this tessellation of squares and parallelograms the squares are solid pieces, hinged at their corners, and the parallelograms are merely empty space. Will the hinged squares be movable? What will their extreme positions be?

(If shuffling cardboard squares around with your fingers seems too tricky, they can be cut out with tabs, as in the figure, and split pins pushed upside down through the overlapping vertices.)

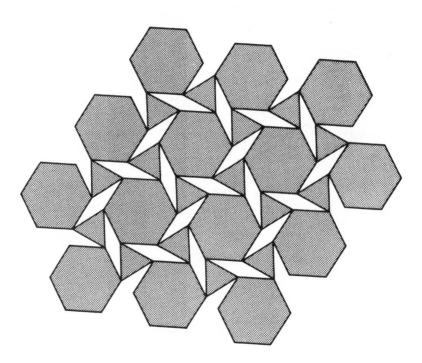

Can this tessellation of hexagons and triangles be hinged as in problem 4?

6

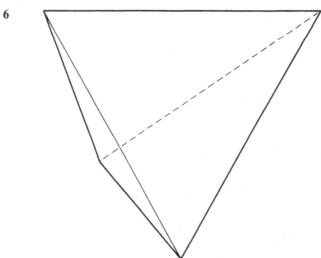

Instead of joining some of the vertices of a dodecagon to form squares, join some of the vertices of a cube to form a tetrahedron, a regular solid with four triangular faces.

3

Crystals

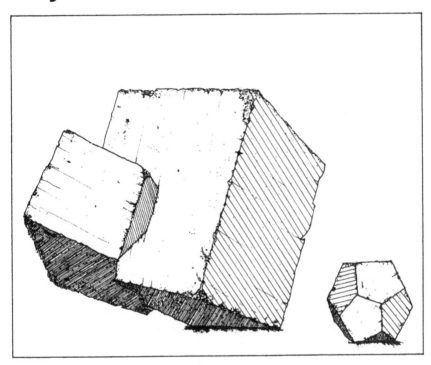

For millions upon millions of years, mineral crystals have displayed mathematical beauty and symmetry in nature. Eons before reptiles and birds were laying their smoothly oval eggs, even before primitive plants were displaying a simple symmetry in their leaves, or primitive organisms in the sea created symmetrical shells and skeletons, mineral crystals were forming deep in the earth.

Ironically, only *Homo sapiens*, a recent arrival in the animal kingdom, has as far as we can tell the ability to admire and appreciate their beauty. No wonder early scientists decided that mineral crystals grew within the earth, just as they supposed that living organisms were spontaneously generated in dirt. How else could flies and bugs suddenly appear out of apparently dead matter? How else could crystals acquire their symmetry, except by growth according to some living principle?

We have no cause to laugh at them. Crystals are still described as 'growing' in a solution. Our advantage is that we know, more or less, how they grow, and how they achieve their regularity despite never having been alive.

Both crystals in the picture are of iron pyrites, or iron sulphide, FeS_2. The crystal on the left is in the form of a pair of interpenetrating cubes. No need to invent the cube, if you can find such cubes hidden in the ground!

The second crystal would be described by a mineralogist as a pyritohedron. Mathematicians would simply call it a dodecahedron, but that is an ambiguous description as will be seen.

Such crystals of pyrites are still found in Italy, where the early Greeks established colonies. The early Greek philosophers may well have seen them there, and hence 'discovered' the idea of a perfect 12-faced solid, with perfect pentagonal faces. The dodecahedral shape was also known to the Celts. More than two dozen dodecahedra of Celtic origin have been preserved. The dodecahedron, the octahedron, the cube and the tetrahedron were the four regular solids known to Plato. The icosahedron which has 20 triangular faces was a later discovery.

All crystals belong to one of several classes or systems. It is no surprise that the cube belongs to the cubic system. So also, however, do octahedral crystals with this form.

How come? Where is the 'cubeness' of the octahedron? This also is revealed naturally, in actual mineral crystals.

Galena is a sulphide of lead. Like iron pyrites its crystals are often cubical. However, the typical crystal of galena will not be a perfect cube. The chances are that its corners will be missing. Perhaps a little, perhaps a lot. Actually, some corners will be missing more than others, but we can correct this, mathematically, to produce this sequence of idealised crystals of galena.

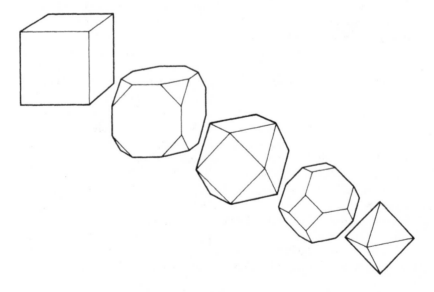

The first crystal is a cube. The second is 'obviously' a cube with its corners cut off, is it not?

Slice off rather more and the 'cubeness' is distinctly less clear in the third crystal which is a kind of half way house, a regular solid called the cuboctahedron. The fourth crystal looks more like an octahedron with its corners removed. The fifth crystal is a perfect octahedron.

The transformation is smooth, subtle and, as it were, symmetrical. We might just as well have started with the octahedron and sliced its corners to finish with a cube.

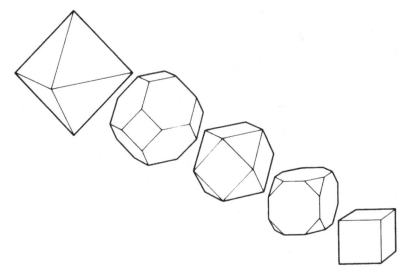

What could be more symmetrical? Yet, if that is the case, why is it not called the octahedral system?

The outer form of the crystal only suggests how the molecules of the mineral fit together in a regular structure. Hauy (1743–1826) 'the father of crystallography', explained how a regular structure of tiny cubes could produce a variety of outward forms.

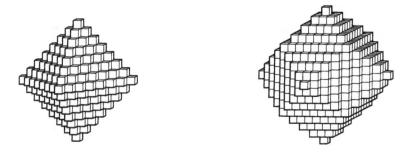

This octahedron has stepped faces, and 'in theory' such a surface ought to be rough, but the actual faces would be so unimaginably large compared with their building blocks that the faces would appear to be quite smooth.

There are other ways of stacking cubes at a sub-microscopic level. For example, a pyramid may be built on each face of an initial cube as in the second diagram.

The result is literally a dodecahedron, and is often called so by mineralogists, because it has twelve faces, 'dodeca' meaning twelve in Greek. Each of the twelve original edges of the cube has been replaced by a diamond-shaped face.

Mathematicians and sometimes mineralogists, call it a rhombic dodecahedron, to distinguish it from the regular pentagonal dodecahedron. Its faces are not themselves as regular as they might be. A regular four-sided polygon should be a square, not a diamond.

Crystals of the cubic system are not limited to these shapes. Indeed, there are no less than fifteen crystals of cubic form, and these include the tetrahedron, or regular triangular pyramid.

The tetrahedron? Where is there any cubicalness in that shape? In many ways!

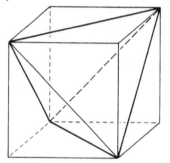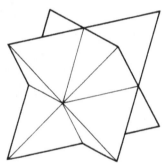

Joining every other corner of a cube, forms a tetrahedron. The corners missed out first time, form a second tetrahedron.

It can now be no surprise that the regular dodecahedron is related to the cube also. The relationship is not so obvious, but that is merely another way of saying that it is even more beautiful.

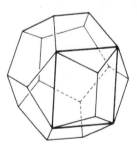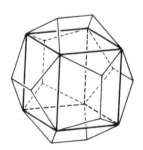

Examination of an actual dodecahedron reveals that the four vertices marked in the figure form a square. Turning the dodecahedron over in our hands, we can find the six faces of a cube.

How many cubes can be fitted into one dodecahedron in this manner? Looking carefully, it is apparent that each edge of the cube is one diagonal of a pentagonal face. But each face has five diagonals, which suggests that a maximum of five cubes might be fitted in without duplication and this indeed is so.

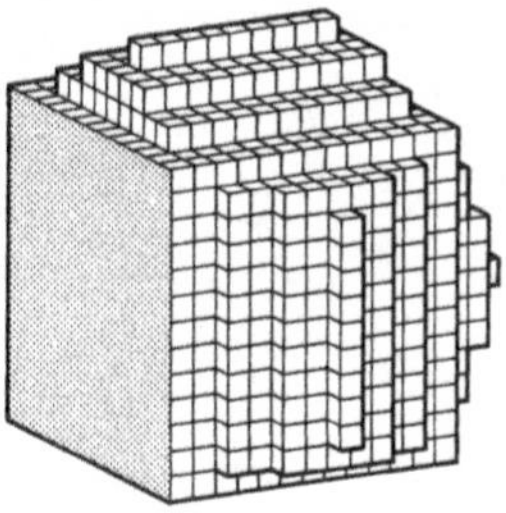

The ingenious Hauy was also able to construct the dodecahedron by building pyramids onto his basic cube. For the dodecahedron, however, the pyramids are ingeniously unsymmetrical, rising up to a line, like the ridge of a tent, rather than to a point, just as in the diagram on p. 32. How smoothly one symmetrical form is transformed into the other!

Problems

1 Since a regular tetrahedron is one of the cubic class of crystals it should be possible to find in a regular tetrahedron three axes symmetrically placed at right angles to each other. Where are they?

2 Find an octahedron in a regular tetrahedron.

3 Compare the volume of a cube with the volume of one of the tetrahedra inscribed in it. (It may help to know that the volume of a pyramid is one third of the volume of the prism on the same base and with the same height.)

4 Compare the volumes of a regular tetrahedron and a regular octahedron which have edges of the same length, by using the result of problem 2.

5 Starting with a packing of cubes, construct a tetrahedron in each cube so that any point which is the corner of one tetrahedron is the corner of eight tetrahedra.

    What shape are the gaps between the tetrahedra?

6 In how many ways can a regular tetrahedron and an octahedron be sliced to show a square and hexagonal cross-section respectively?

4

Circles and ellipses

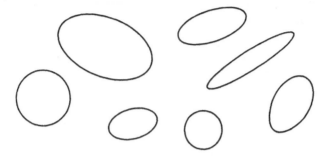

Which of these ovals is a circle? Which of them **could be** a circle seen at an angle? Only two of them are circles, but only one of them could not be a circle seen at an angle. The oval lower right is 'obviously' wrong because it is not sufficiently symmetrical.

When a circle is seen obliquely, it appears to be elliptical, in the shape of an ellipse. Our eyes are so accustomed to this distortion or transformation of the circle, that when we look round a room we correctly interpret most of the ellipses we actually see, as circles. The edge of a lampshade, the rim of a pot, a circular rug, mats on the table, a fruit dish . . .

However, the oval of light reflected by the circular mirror on the wall is a genuine ellipse and not a circle.

Ellipses are typical of many mathematical objects. They turn up all over the place, in many different ways, at first sight unconnected with each other.

An ellipse is, most simply, a squashed circle. Take a circle and draw a diameter through it. Mark some points on the circle. Now bring those points closer to the line. For simplicity, let's bring each point twice as close to the line.

Point P will become point Q, and so on. The circle has now been squashed so that it is only half of its former height. The result is an ellipse.

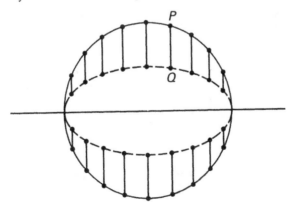

35

This method of drawing an ellipse is easy to understand, but it is tedious to perform.

It is more elegant to adapt the usual method of drawing a circle.

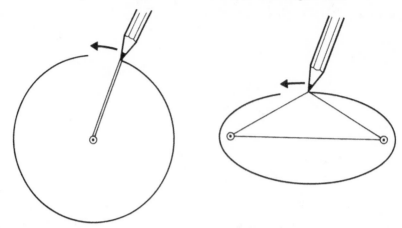

If you lack a pair of compasses, you can draw a circle with a pin and string, like this. To draw an ellipse, replace the single pin by two pins, a distance apart. The further the pins are apart, the more squashed the ellipse will be.

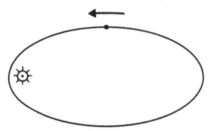

The astronomer Kepler first proposed that the planets travel round the sun in elliptical orbits. They do, to a very good approximation, and the sun is placed at a point corresponding to one of the pins.

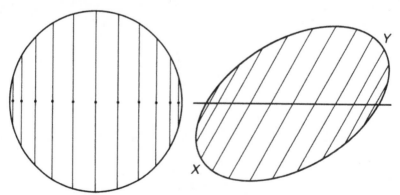

A circle can also lose height if a diameter and the lines parallel to it keel over like this. The circle has not simply been squashed, at least not vertically.

The narrowest points of this oval are stuck out at X and Y. Yet this shape is still an ellipse!

What is the opposite of being squashed? One opposite is to be stretched, and stretching a circle also produces an ellipse. Alternatively, we can think of an ellipse being stretched to get back to the original circle. Or an ellipse can be squashed to produce a circle, or another ellipse.

The initial circle has been squashed to half its height to make an ellipse, which has been stretched by 50% to make a longer ellipse, which has been stretched by a factor 3 vertically in order to get back again to a circle.

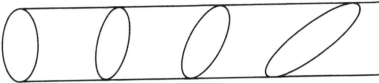

When a cylinder with a circular cross-section is sliced, the cross-section is, as it were, stretched out along the cylinder. Its shape becomes an ellipse. Any section of the cylinder is circular or elliptical. Every ellipse has the same diameter as the initial circle in one direction, but its diameter in the direction at right angles can be as long as we please.

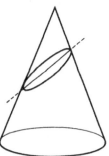

This is a cone which, like the cylinder, has been sliced. What shape will the slice expose?

First thought might suggest that shape will be widest at the bottom, where the cone is widest, and narrower near the top, towards the point.

Or perhaps, on second thoughts, the top of the slice will be wider because that is where the slice cuts the edge at the largest angle? At the bottom it cuts the cone at a much smaller angle and there is no doubt that if you make several cuts at the same point on a cone, the smaller angle will produce the narrower section.

Curiously and extraordinarily both these arguments have exactly equal force, as it were. They cancel each other out! The slice is actually completely symmetrical. It is another ellipse.

A connection with slices of cylinders is easy to see. Any cylinder can be thought of as a cone whose point (whose vertex) has been stretched an infinite distance, so that the sides of the cone are now parallel to each other, and parallel slices are identical in size and shape.

Is there also a connection with the pin and string method of drawing ellipses? Of course there is! It was discovered by the ancient Greek mathematicians who first investigated ellipses as slices of cones more than two thousand years ago, and long before Kepler discovered that he needed ellipses to explain the movement of the planets.

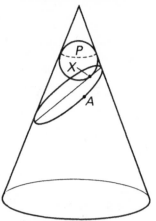

They imagined a cone, sliced as before, and a sphere placed between the plane cut and the point of the cone, so that it just touched the plane cut at one point, marked X, and also touched the cone all round the circle marked P.

Consider a point on the ellipse. Call it A. A straight line joining A to X will touch the small sphere. So will a line going straight up the side of the cone from A to meet the circle P. What is more, these two lines will be the same length because **any** two lines from one point which touch the same sphere will be of the same length.

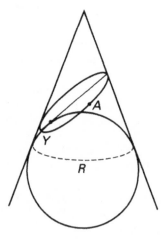

Next they considered a second, matching sphere, touching the cutting plane at Y and touching the cone all round the circle marked R.

Just as with the small sphere, the straight line joining A to Y and the line going straight down the side of the cone from A to meet the circle R, are of the same length.

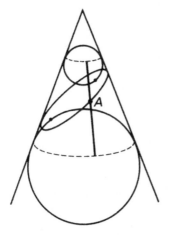

Putting the two ideas together, the total length of $AX + AY$ is always equal to the length of the line which starts at the circle P immediately above A, and goes straight down through A to meet the circle R.

Because this straight line is always of the same length (it is just the distance between the two circles), $AX + AY$ is always of the same length, which is why the pin method for drawing ellipses works!

Problems

1

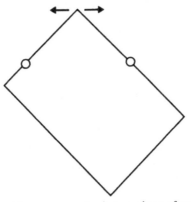

Press two drawing pins or map pins into a piece of paper and place an ordinary sheet of paper so that it slides between them, as in this diagram. What is the path of the corner of the paper?

2

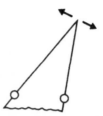

Press two pins into a sheet of paper, as in the last problem, but move between them one of the smaller angles of a set square, like this. What shape does the moving corner trace out?

3

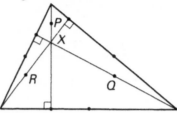

In this triangle, which has not been chosen to be any special shape, 3 lines called altitudes have been drawn from each corner at right-angles to the opposite side. They meet in one point, marked X. P, Q and R are half way between that point and the corners.

What is special about the middle points of the sides of the triangle, the points where the altitudes meet the sides, and P, Q and R?

4

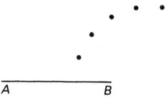

A   B

Mark two points on a piece of paper, call them A and B, and mark as many points as possible which are twice as far from A as they are from B.

What is special about all these points?

5 According to the poet Virgil, Aeneas during his wanderings visited Queen Dido of Carthage, who offered him as much land as he could encompass with the hide of an ox. Assuming that he had the foresight to cut the hide into a long narrow strip, and that he exploited the straight shoreline as one boundary of his domain, how should he go about enclosing as much land as possible?

6

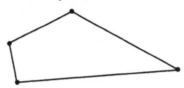

This quadrilateral is hinged at the corners so that its shape and its area can vary. When will its area be a maximum?

Skeleton and structure

What have the seal, the bat, the horse, and a human being in common? At first sight very little. However, first sight as so often is misleading. Their outer forms, habitats and habits are indeed very different but inside is another story. They are all mammals and therefore their internal organs are similar though their outward forms are distinguished. In particular their skeletons are astonishingly alike.

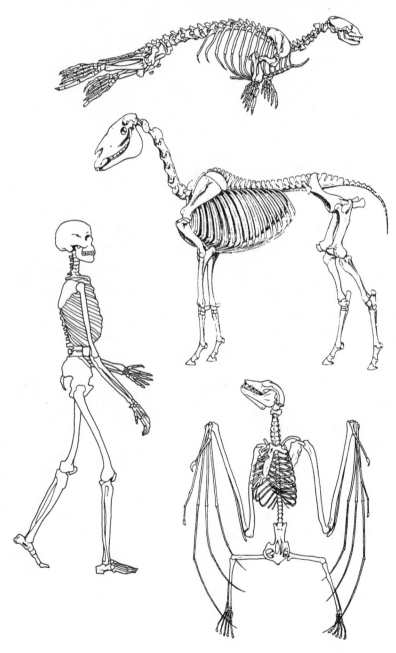

The seal has a tail, it is true, and human beings only have a small bone at the base of the spine called the cocyx, but every skeleton has a backbone with ribs attached, though the number of ribs varies. Even more extraordinary, the bat's wing, the seal's flipper and the human arm are more similar than different under the surface.

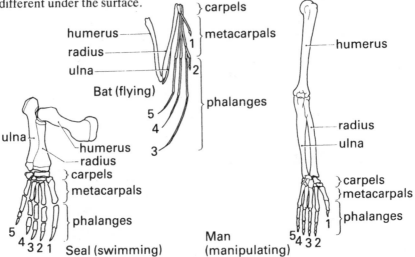

The person in the photograph is able to twist their lower arm from the elbow, by rather more than 180 degrees. How is this possible? The upper arm cannot be rotated in the same way from the shoulder. The difference lies in the bones of the arm. There is only one bone in the upper arm, but in the lower arm there are two. When the lower arm rotates, its two bones twist round each other, as it were. Neither bone twists more than very slightly in its socket. The same masterpiece of design is found in all these skeletons. Since it is also found in the dinosaurs, it is at least two million years old.

How similar are these two wallpaper patterns? Superficially they may seem very different, in style, pattern, texture and feel.

What happens, however, if instead of highlighting the details of the decorative features, the arrangement of these features is brought out?

The same point of each decorative motif has been highlighted and the decorations themselves are fading into the background. The similarity between the patterns has increased considerably.

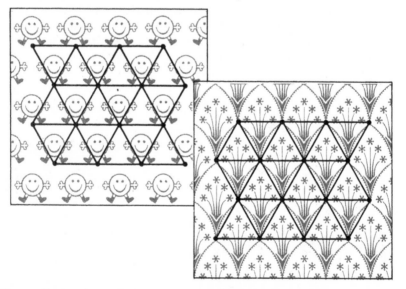

To emphasise the underlying skeleton more strongly, the decoration has been faded further, and the corresponding points in each motif have been joined by straight lines. It would hardly be an exaggeration to say that these basic skeletons are almost identical, and that the original designs have the same basic structure.

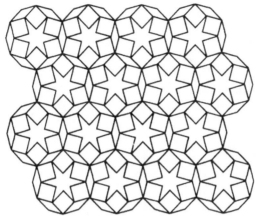

Islam forbids the use of human images in art and so Muslim artists have been forced to develop more abstract forms of decoration. This pattern is a very simple example, which nevertheless is very rich.

An obvious way to find a structure for it is to mark and join the centres of the six-pointed stars. It can then be seen as composed of the unit on the left repeated in all directions. However, this does not seem the most natural way to structure this pattern. It is surely simpler to see it as repetition of the second unit, where the corresponding points are the vertices of regular hexagons. These small rings of squares certainly hit the eye strongly, but looking more closely it might seem that it is the larger rings of squares on the outside of the very same hexagons which are 'simplest'.

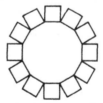

Since the number of squares in the pattern has not increased, each square must appear in two units if the unit is twice as large. In this case, each square is a member of two of the large rings.

It is not necessary, and certainly not easy, to argue about which way of looking at the pattern is simplest. The important fact is that whether through a grid of triangles, or hexagons, this pattern can be perceived as constructed in a simple manner which we could easily copy, vary or develop.

It is a curious and mysterious feature of our brains, that they often take a long time to work out, in so many words and images, what they can see as an underlying pattern in no time at all. No one needs to analyse the underlying skeleton of either of the wallpaper designs to appreciate that they are patterns. It is only when we choose to draw one, that we realise that a different and deeper understanding which we can put into words or images, is required.

Our brains are just as clever at recognising non-visual patterns in music or literature. How easy it is to recognise the rhythm of rock and roll, or reggae, or a cha-cha-cha! Yet how many readers could explain in so many words the difference between them?

It is not difficult to recognise a limerick from the rhythm, even though not all limericks have identical patterns of stress and metre:

> a DA da da DA da da DA,
> plonk PLONK plonk plonk PLONK plonk plonk PLONK:
> When he CRIED where's the MAID,
> she rePLIED, I'm AFRAID
> a PLONkety PLONKety PLONK

Writers and poets in idle moments write parodies which closely follow the structure of the original, and maybe its tone and feeling, and yet 'take the Mickey'.

It is probably no coincidence that Lewis Carrol, alias Charles Ludwig Dodgson, a mathematics tutor at Christ Church, Oxford, was a great parodist but a complete failure as a Serious Victorian Poet. Trust a mathematician to emphasise the pattern at the expense of the everyday meaning.

Problems

1

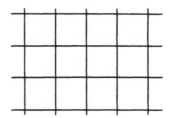

Construct a tessellation of small squares and octagons by 'slicing the corners' of the tessellation of squares.

2

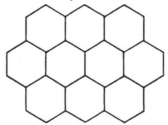

Make a new tessellation by spreading these hexagons apart a little, and filling the continuous gap between them with squares and equilateral triangles.

3 How many semi-regular tessellations are there whose tiles are regular hexagons and equilateral triangles? Semi-regular means that the same selection of tiles surrounds every vertex in the same order.

4 If every other vertex of a honeycomb tessellation of hexagons is picked out, what tessellation will they make?

5

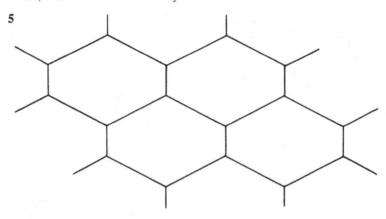

This tessellation of elongated hexagons can form a tessellation of identical pentagons, as problem 1 of Chapter 2 explained.

Here is another simple property. Mark every other vertex of each hexagon. What tessellation do they make?

6

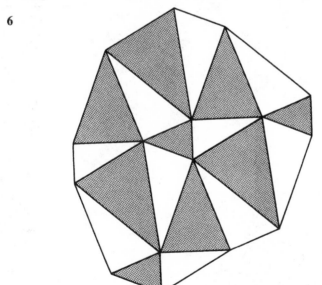

This tessellation is made from two sets of triangles. The shaded triangles are all similar to each other and so are the plain triangles.

Can this tessellation be seen as a tessellation of irregular hexagons? Of two different pentagons? Of three different quadrilaterals?

Taking the wider view

This is an arbitrary triangle with an equilateral triangle fixed to each edge. At first sight there may seem to be nothing special about it.

However, if we mark the centres of the equilateral triangles, and join them, the result is, surprisingly, another equilateral triangle. This is no accident. Try the experiment starting with any shape of triangle you choose and it will always work, a result that is called Napoleon's theorem after the French general who is supposed to have discovered it.

To discover pattern and symmetry where there appears to be neither is always delightful and intriguing. It also suggests that we are not seeing all there is to be seen in the original diagram. There must certainly be another way of looking at it which will be more symmetrical from the start, and therefore make the conclusion more natural.

One idea, which sometimes solves geometrical problems, is to think of the original figure as a part of a tessellation, a tiling of the plane which could repeat in a regular pattern for ever. To see if this idea will work, consider the large equilateral triangle as the 'centre' of the diagram, and repeat the other three triangles on its other two sides.

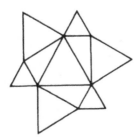

So far, so good, but now comes the first test. Can the narrow gaps round the edge be filled in a regular manner? Yes, they can, by fitting one more of the original triangles into each.

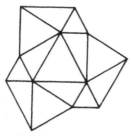

The diagram is already looking much more symmetrical, but the symmetry can be increased still further by continuing to expand outwards.

49

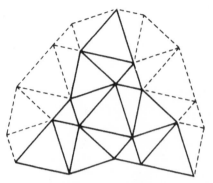

The large equilateral triangle is now being repeated for the first time, and it is not difficult to see that the pattern will go on for ever in all directions.

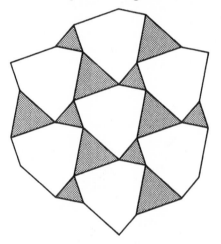

By shading some triangles and omitting some lines, the symmetry is even clearer.

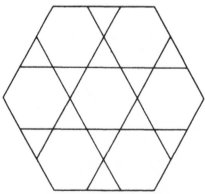

This diagram can be seen as a distortion of a well-known tessellation of regular hexagons and equilateral triangles. The symmetry of this tessellation is easy to see, but the symmetry of our tessellation is almost as clear.

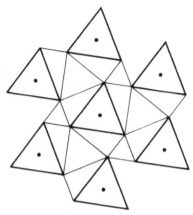

The shapes of the hexagons have been distorted but their arrangement is as regular as ever. Looking at the middling equilateral triangles and the large equilateral triangles that surround them, we can see a similar kind of symmetry.

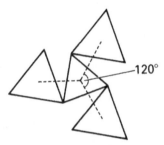

Joining the centres we have three spokes, with angles of 120° between them. The same symmetry appears if, for example, the centre of a small equilateral triangle is joined to the centres of its nearest middle-sized triangles, or any other combination of triangles is used.

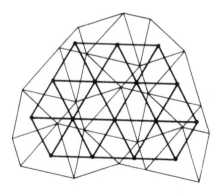

Covering the whole tessellation in a network of spokes emphasises that the skeleton of this tiling is indeed identical to the common tiling of equilateral triangles. In particular, we may notice almost in passing, that the centres of any three equilateral triangles surrounding one of the original triangles, form also another equilateral triangle.

Although the figure is so very symmetrical it may seem that our explanation has been unsymmetrical. Why have we concentrated on the large and the middling equilateral triangles? Surely, it should be just as simple to see the answer by looking, for example, at the smallest and middling equilateral triangles?

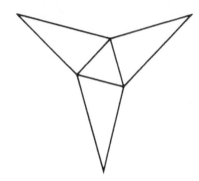

Indeed we can do so, though it may not be quite so easy to see, because our brains are not so accustomed, nor indeed so clever, at seeing star-shaped hexagons. However, we can see that this tessellation has indeed the same brilliant symmetry of the diagram on page 49.

This tessellation solves the original problem, but it is by no means the only way of looking at it anew, merely one of the simplest. Also, it fails to answer other questions which might spring to mind, such as, why did the original extra triangles have to be equilateral? An equilateral triangle is symmetrical in itself, but the relationship between the original arbitrary triangle and the equilateral triangles is quite unsymmetrical. It is like comparing a perfect circle with an eccentric ellipse. Why cannot triangles of any shape be fixed to the original triangle?

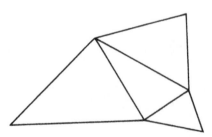

This diagram looks like our original figure, except that the three added triangles are not equilateral, merely similar to each other in shape. Depending on how you look at it, such irregular triangles either do not have centres at all, or they have several different 'centres'. No matter, we will simply join corresponding points, as usual.

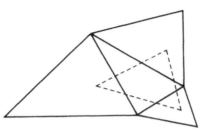

What shape is the triangle they form? The same shape as the three original triangles. This is one generalisation of the original idea. Can this experimental fact be explained by using a tessellation? Yes, if the meaning of the word tessellation is stretched a bit.

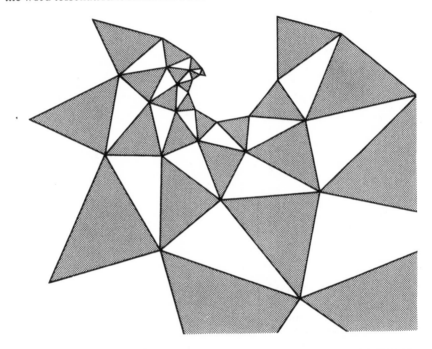

To make this tessellation come alive, imagine that there is a kind of black hole in the centre of the paper. As the triangles spiral round it they are sucked towards its centre, getting smaller and smaller but preserving their shape.

Concentrate on the shaded triangles. They are all the same shape and they are all in the same, constant relationship to each other. Space itself is curved by the influence of the black hole at its centre, and in particular any shaded triangle can move smoothly into the position of any other shaded triangle, without changing 'its shape. Picking three corresponding points from three linked shaded triangles is just one way of describing the position of a shaded triangle as it moves across the spiral surface.

What about the unshaded triangles? They too move about the spiral surface without distortion, apart from their change of size. Like positive and negative of the same photograph, each set of triangles shares the property of the spiral.

Problems

1 Conclude that the angles of an arbitrary triangle sum to 180° by constructing a tessellation of identical triangles.

2 Conclude that the angles of an arbitrary quadrilateral sum to 360° by constructing a tessellation of identical quadrilaterals.

3 This method of dissecting two arbitrary squares to make one square was discovered by Abul Wafa (940–997). The larger square is cut into four quarters which are placed round the smaller square. When four cuts are made along the dotted lines, the bits cut off miraculously fill the inside spaces to complete the single square.

Show how Abul Wafa might have discovered this method, by making a tessellation out of repetitions of the diagram.

4 Desargues' theorem says that if two triangles are 'in perspective' from a point, then the intersections of corresponding sides lie on a straight line.

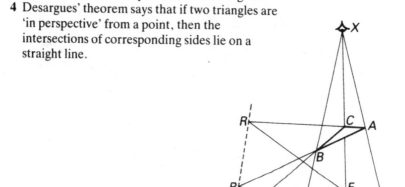

This means in this diagram that because their corresponding vertices appear to be in line to a viewer at the point X, the points where pairs of corresponding sides meet, P, Q and R, lie on a straight line, as shown.

The problem is to 'see' this flat 2-dimensional diagram as a 3-dimensional arrangement of points, lines and planes, so that Desargues' theorem becomes obvious.

5 Continue the sequence: 0   1   12   63   208   525   1116   . . . by looking at it as a part of this pattern:

0		1		12		63		208		525		1116
	1		11		51		145		317		591	
		10		40		94		172		274		
			30		54		78		102			
				24		24		24				
					0		0					

6 Reconstruct the missing term in this sequence by looking at it as part of the kind of pattern which appears in problem 5.

6   15   40   –   162   271   420   615   862

54

7

Invariants

An experiment with an old hollow rubber ball and a knife or scissors will suggest, correctly, that the surface of a sphere cannot be flattened into a sheet without some distortion. Indeed, even a small piece of the surface, such as one panel of a football, will be distorted a little if forced flat. It will either be wrinkled, or stretched, or maybe torn as well.

The entire surface of the earth must also be distorted before it can be displayed flat. Cartographers therefore face a tricky problem. Some features of their maps are going to be wrong. How many? What features can possibly be preserved? Although the news for mapmakers is grim, it could be worse. It turns out that they have a choice of which features they will preserve. The maps below illustrate two possibilities.

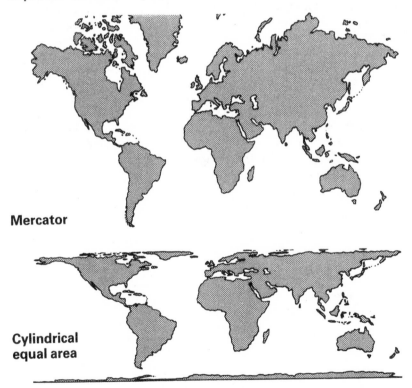

Mercator

Cylindrical equal area

The Mercator projection was introduced in 1567. It is probably the best known of all projections and is widely used in atlases and wall charts. It has the immense advantage for navigators that courses of constant bearing, called loxodromes or rhumb lines, become straight lines on the map. The navigator can draw a straight line from the ship's present position to its destination, measure the bearing, set the automatic compass, and go to sleep.

(This does not entirely solve the navigator's problems. The shortest distance between two points at sea is an arc of a great circle which is not a course of constant bearing. However, a great circle course can easily be closely approximated by several shorter courses of constant bearing.)

Mercator's projection inevitably has disadvantages also. The comparison of areas is especially difficult, because the lines of latitude move further apart as they approach the poles. A well-known absurdity concerns the relative sizes of Greenland and South America. On a Mercator map, Greenland appears rather the larger of the two. In actual fact it is only about one tenth the area of the South American continent. Loxodromes are an invariant of the transformation which produces Mercator's projection. Area is not invariant.

The cylindrical equal area projection does preserve areas, though it changes their shape. Loxodromes are no longer invariant and this cylindrical projection would be a navigator's nightmare. Greenland now appears accurately small, while South America which straddles the equator is little changed. Europe itself is somewhat reduced in size because on Mercator it is far enough from the equator to be slightly enlarged. In contrast Africa which also crosses the equator is about the same size, and so appears relatively larger than Europe, as it should do. The atlases of my and many readers' childhoods made Europe larger than it is, and also more important by placing it in the centre of the map.

Mathematicians are continually searching for ways to transform one object into another, or one problem into another. If some features remain their old familiar selves, so much the simpler.

One of the simplest invariants in all of elementary mathematics is area. When a plane shape is cut up and reassembled in a different way the area does not change. This is how mathematicians have always calculated the areas of triangles and parallelograms and other simple shapes.

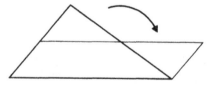

Here is one way to turn a triangle, any triangle, into a parallelogram. The area has not changed, it has merely been rearranged.

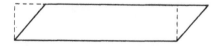

The parallelogram can in turn be transformed into a rectangle by dissection. Here a right-angled triangle is cut off one end and placed at the other.

The original parallelogram and the new rectangle have the same area.

The easiest ways to design variations on well-known patterns depend on transformations which leave the basic structure alone.

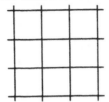

All the symmetry of this square pattern will remain invariant if at each vertex we grow identical small squares.

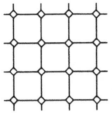

These squares can grow larger, and larger, until they nearly meet each other, or even overlap each other. The symmetry remains.

More subtle, hidden and mysterious, but just as powerful, are the relationships between the numbers of faces, corners and edges of polyhedra.

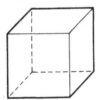

Here are a cube, and a cube with a missing corner. The cube has 6 faces, 12 edges and 8 vertices. The number of edges almost equals the number of vertices and faces together: in fact 12 + 2 = 6 + 8.

By slicing off a corner, an extra face appears, also three extra edges and two extra vertices. The edges still fall short of the faces and vertices by 2: 15 + 2 = 7 + 10.

Is this difference of 2, an invariant also? Slicing a corner where three faces meet will certainly not change it. 1 extra face and 2 extra vertices will balance exactly the 3 extra edges.

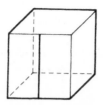

How about dividing one face into two? The edges increase by 3, because the edges joined by the new line are each bisected. There is 1 more face, and 2 new vertices. No change!

How about slicing parallel to an edge? The edges have increased by 3. There is 1 extra face and 2 extra vertices. No change again!

What about other shapes? A tetrahedron is much like a cube with four corners sliced off. Indeed it **is** a cube with four corners completely missing, but let us count anyway.

The tetrahedron has 6 edges, 4 vertices and 4 faces. The edges are still 2 short.

It does seem that slicing off corners and edges, or dividing the faces, makes no effective difference. What about gluing two cubes together?

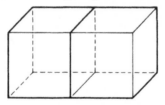

Here are two cubes face to face. Two faces have disappeared completely, and the 8 edges and 8 vertices which surrounded those faces have been reduced to 4 edges and 4 vertices. The totals are now 20 edges, 10 faces and 12 vertices. The pattern remains, which is not perhaps so surprising, since on reflection the two cubes face to face look much like a single cube with four faces divided into two parts each, and we already know what happens when one face is bisected.

Have we now considered all the possibilities? By no means! Suppose we persevere with gluing cubes together, but not in a straight line.

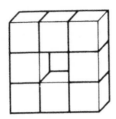

These eight cubes make a ring, which we might hope would provide something new because as the ring of cubes is completed we are losing an unusual number of edges, faces and vertices. We are not disappointed. This square doughnut ring has 32 faces, 32 vertices, and 64 edges. There is no shortfall in the number of edges, and the value of 'faces + vertices – edges' is not invariant for all polyhedra. At the very least it can have either of the values 2 and 0.

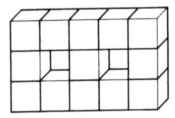

Are these the only possibilities? When a trick works once, it is always tempting to try it again. This polyhedron is pierced by two holes. What is its total 'faces + vertices – edges'?

Problems

1

It might seem completely and utterly obvious that if you draw a closed curve, as in this figure, then it divides whatever you are drawing on into two parts, an inside and an outside. Prove that this is **not** a feature of all surfaces by finding a surface for which it is not true.

2 Take a slice off a circle, and join the ends of the cut to any other point on the circle, such as T. Whatever point you choose for T, the final figure will have a particular feature. What is it?

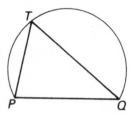

3

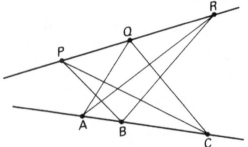

Here are two arbitrarily chosen lines. Three arbitrary, distinct points have been chosen on each line. These points are labelled A, B, C and P, Q, R. Each point has been joined to the two points on the other line which are **not** 'directly opposite' to it.

The finished diagram has a feature which is independent of the positions of the original 6 points. What is it?

4 What common operation leaves the value of a fraction invariant?

5 1 1 2 3 5 8 13 21 34 55 89 144 233 377 . . .

This is called the Fibonacci sequence. Each number is the sum of the two previous numbers. The ratios of successive terms rapidly get closer and closer to the Golden Ratio, which is equal to 1.681033 . . .

For example, $3/2 = 1.5$, $21/13 = 1.61538 . . .$,
$$377/233 = 1.618025 . . .$$

What is the limiting ratio when the series starts with a different pair of numbers, for example 2 and 9:

    2 9 11 20 31 51 . . . ?

6 Take an arbitrary triangle ABC, and draw the line from A to BC, at right angles to BC.

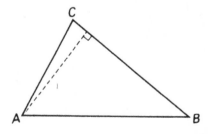

Add the similar lines, called altitudes, from B to CA and C to AB. Whatever triangle you start with, the final figure will always have one special feature. What is it?

Special cases

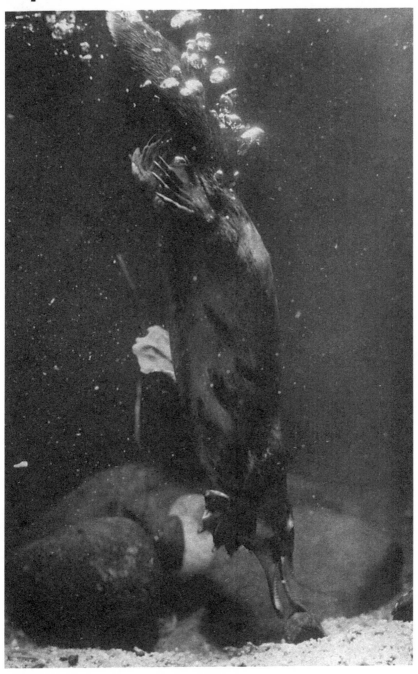

Every feature of a triangle, except its position, is fixed and can be calculated when the lengths of the sides of the triangle are known. In particular its area can be calculated by using this formula:

$$\text{Area} = \sqrt{s(s-a)(s-b)(s-c)},$$

where a, b and c are the lengths of the three sides, and s is one half the sum of the sides. For example, if the sides were 13 cm, 14 cm and 15 cm then half their sum would be 21 cm and the area would be $\sqrt{21(21-13)(21-14)(21-15)}$ which is $\sqrt{(21 \times 8 \times 7 \times 6)}$ or 84. (This is a rare case of a triangle with integral sides whose area is also integral. Usually the area will be irrational.)

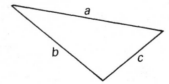

This formula is easy enough to use and as expected it is very balanced. Inevitably, the lengths of the sides, a, b and c, appear in it symmetrically. The factor s at the start, however, is less symmetrical. Why is it there?

To answer this question by starting from the formula would be extraordinarily difficult, if not impossible. However, the answer is easy to see from another even more beautiful formula which was known to the Indian mathematician Brahmagupta.

Generally, it is not possible to calculate the area of a quadrilateral from the lengths of its sides because a quadrilateral, unlike a triangle, is not naturally rigid and as the sides rotate about each other the area will vary.

However, we can fix the shape and the area of any quadrilateral by insisting that its vertices lie on a circle. It is then called a cyclic quadrilateral.

Brahmagupta's formula is

$$\text{Area} = \sqrt{(s-a)(s-b)(s-c)(s-d)},$$

where a, b, c and d are the lengths of the four sides, and s, as before, is one half of their sum. Notice that the formula is now completely symmetrical. Notice also that it is remarkably similar to the formula for the area of the triangle. How can we 'move' from one to the other?

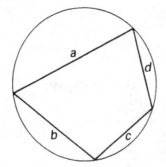 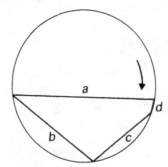

The answer is simple. We think of the triangle as a special case of cyclic quadrilateral by imagining that one side of the quadrilateral gets shorter and shorter, until 'its length is zero'. At that point (no pun intended!) the cyclic quadrilateral will have become a triangle and the term s appears where Brahmagupta's formula has the factor $(s - 0)$. Since any triangle, without exception, can be drawn with its vertices on a circle, we can think of any triangle as a cyclic quadrilateral, and calculate its area by Brahmagupta's formula.

How important and useful is this idea of a triangle as a cyclic quadrilateral? In most circumstances the answer is, not useful at all. For example, Ptolemy's theorem states that in any cyclic quadrilateral the product of the diagonals is equal to the sum of the products of the opposite pairs of sides.

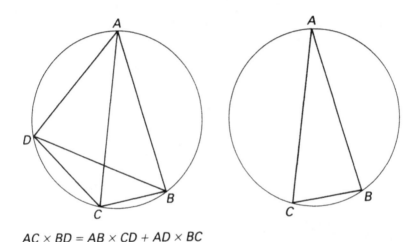

$AC \times BD = AB \times CD + AD \times BC$

Ptolemy's theorem also has a special case in which one side is of zero length. It will then tell us something about any triangle, but what will this message be? Imagine the length of one side decreasing to zero, so that point D becomes point C and the length of DC is reduced to zero.

We end up with a triangle ABC in which, so Ptolemy's theorem tells us, $AC \times BC = BC \times AC$. How obvious! How uninteresting!

Brahmagupta's formula and Ptolemy's theorem, applied to a triangle, illustrate two extreme possibilities. The special case of Brahmagupta's formula is extremely powerful and useful, while the special case of Ptolemy's theorem is trivial and absurd.

When a special case is created by making a length equal to zero, or making two points coincide, there is always a danger that the result will be trivial. Indeed, sooner or later the result must be trivial, because there will be no points or lines left to make an interesting figure!

A cyclic quadrilateral is itself a special case, but of great interest. The extra condition that its vertices lie on a circle gives it a wealth of properties that are completely lacking in most quadrilaterals. As well as Brahmagupta's and Ptolemy's results there are, for example, the properties of its opposite angles discussed in Chapter 1, and the special feature of this figure.

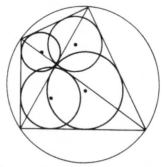

A circle has been inscribed in each of the four triangles formed by the quadrilateral and its diagonals. The centres of the circles form a rectangle.

The moral seems to be, in mathematics as in art and everyday life, that too few conditions provide too little structure while too many conditions squeeze out all the interesting possibilities!

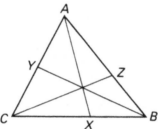

Ceva's theorem says that if three points are chosen on the side of any triangle so that

$$\frac{AY}{YC} \times \frac{CX}{XB} \times \frac{BZ}{ZA} = 1$$

then the three lines AX, BY and CZ will all go through the same point. A very obvious special case is created when X, Y and Z are the mid-points of the sides, because all three ratios are then equal to 1, and their product is 1 also. What theorem then appears? It is the neat theorem that the medians of a triangle concur.

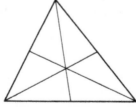

Less obvious is the special case when each side is divided in the ratio of the other two sides. Once again the product of the three ratios is 1, but why is this significant? Because if Z divides AB in the ratio $AC:CB$, then CZ is the line which bisects the angle C and Ceva's theorem now states another familiar theorem that the angle-bisectors of a triangle also pass through one point. This point is the centre of the circle which touches the three sides of the triangle internally.

Problems

1 This is a cyclic quadrilateral with a difference. It is symmetrical about the centre of the circle, which means that the diagonal drawn in the figure goes through the centre of the circle. What can be deduced about the angles at *P* and *Q* in this special case?

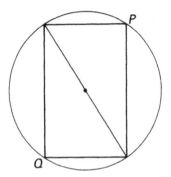

2 Pappus' theorem, the subject of problem 3 of Chapter 7, starts with a pair of straight lines. The conic sections which include parabolas, hyperbolas, ellipses and circles (which are a special type of ellipse) also include, as a very special case, any pair of straight lines.

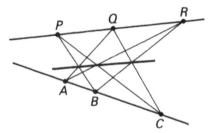

   This is very suggestive. Does Pappus' theorem still work if the 6 points are chosen initially on ANY conic section?

3

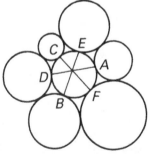

This diagram illustrates the Seven Circle theorem. Six circles each touch the two circles on either side and also touch the inner circle at *A, B, C, D, E* and *F*. The theorem says that the lines *AD, BE* and *CF* concur.

   Imagine that every other circle in the outside chain becomes larger and larger . . . and larger until it becomes a straight line, while the remaining outer circles become somewhat smaller to make room for them. The seven circles continue to touch each other as before.

   What figure and what theorem results?

4

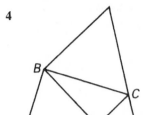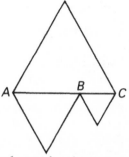

Napoleon's theorem, discussed in Chapter 6, says that the centres of these three equilateral triangles form a fourth equilateral triangle. What happens if the original triangle *ABC* degenerates into a straight line, like this?

5

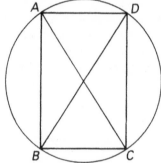

Ptolemy's theorem did not turn into anything interesting when two of the vertices of the quadrilateral moved into coincidence. What happens if they simply form the vertices of a rectangle, as here?

6

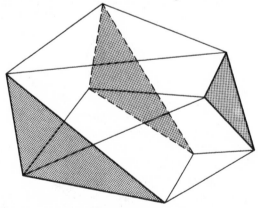

This figure was constructed by drawing two shaded triangles, and then constructing the three similar triangles with the lines joining corresponding corners of the shaded triangles as bases. The third vertices of the three triangles then automatically form a third triangle which is similar to the original shaded triangles. The problem is to say what happens when the three unshaded similar triangles degenerate into straight lines.

9

Summing up

An A1 sheet of card divided into A2, A3, A4, A5 and A6 sizes.

Is it possible to add up a stream of numbers which goes on for ever? Sometimes the answer is obviously 'No':

$$1 + 2 + 3 + 4 + 5 + 6 + 7 + 8 + 9 + 10 + 11 + 12 + 13 + \ldots$$

These numbers get larger and larger and so does their sum, with no end in sight. Perhaps the sum will appear more plausible if the numbers get smaller and smaller:

$$\tfrac{1}{2} + \tfrac{1}{4} + \tfrac{1}{8} + \tfrac{1}{16} + \tfrac{1}{32} + \ldots$$

Does this make a difference? It does, and we can turn this sum into a picture which will make the difference quite clear.

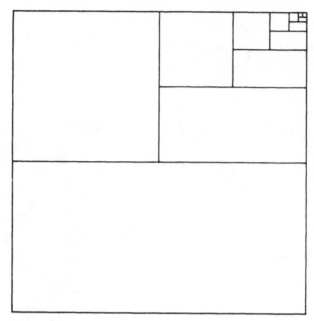

How was this square constructed? We started with an empty square, to represent 1. Then we filled one half, leaving the other half empty.

Next, the empty half was divided into two quarters, and one of these was filled in. Then the empty quarter was divided into two eighths, and one of those was filled in . . . and so on:

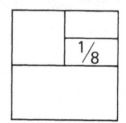

This process can go on for ever . . . and ever . . . and ever, with a little imagination, of course.

Imagination and infinity, however, can be an unreliable, even dangerous combination in mathematics. So mathematicians have learnt to do without the idea of infinity in this instance by concentrating on two features of the picture:

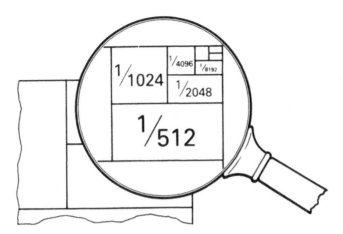

First, no matter how many pieces you fit into the square, you will never quite fill it, so the square will certainly never overflow.

Secondly, by filling in enough pieces, you can certainly get as close as you wish to filling the '1' square completely.

These two conditions are enough to persuade mathematicians that they know exactly what they mean by adding up this endless stream of numbers, without relying on their imagination to peer into an endless 'infinite' future!

This stream of numbers had a very simple pattern; the first number was $\frac{1}{2}$ and each subsequent number was one half of the previous number.

The next series is similar, but a little more complicated: it starts with $\frac{1}{3}$ and each subsequent number is one third of the previous number:

$$\frac{1}{3} + \frac{1}{9} + \frac{1}{27} + \frac{1}{81} + \frac{1}{243} + \dots$$

Can its sum be found in the same way, by filling in a square?

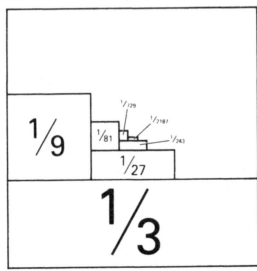

This is one attempt, but the result is nothing like as simple and elegant as our previous attempt.

There is another picture, however, which will do the trick. Start by drawing a strip and dividing it into three thirds.

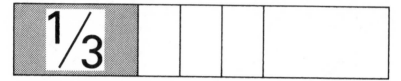

Shade the first third, and divide the next third into three ninths.

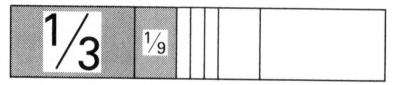

Next, shade the first ninth, and divide the next ninth into three twenty-sevenths.

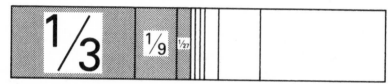

Then shade the first twenty-seventh, and divide the middle twenty-seventh into three eighty-firsts.

And so on! Each step approaches nearer to the half way mark. Indeed we can get as near to the half way mark as we choose; but we shall never pass it, however many steps we take, because we always shade in the first, left-hand piece, which is always slightly to the left of the middle of the strip. This time, the mathematician is satisfied to say that the sum of the endless stream of numbers is **one half**.

So far, so good, but it is a little disappointing that different pictures had to be used for each problem. Will a third new picture be needed to sum this series?

$$\tfrac{1}{4} + \tfrac{1}{16} + \tfrac{1}{64} + \tfrac{1}{256} + \tfrac{1}{1024} + \ldots$$

Is there no general pattern to follow each time? Yes, there is. The sum of the second series is a large hint. If **one** third, plus **one** ninth, plus **one** twenty-seventh . . . sums to **one** half, then **two** thirds plus **two** ninths plus **two** twenty-sevenths . . . should sum to **one**, in which case there is surely an excellent chance that they will fill in the '1' square.

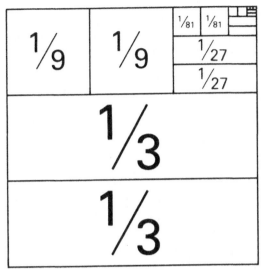

They do. This is the simple pattern they make.

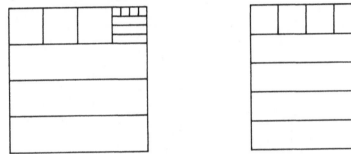

Now this is progress! This same pattern will work for any similar series.

Pictures can be used to sum many other kinds of series. A smart bit of mental arithmetic will show that this series sums to 100 – a suspiciously round number.

$$1 + 3 + 5 + 7 + 9 + 11 + 13 + 15 + 17 + 19$$

Why? This picture supplies one answer.

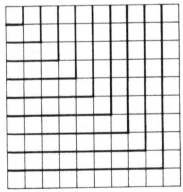

A 10 by 10 square, containing squares, has been divided into L-shaped pieces, plus one single square. Add the sizes of these and you will reach 100. There are other ways to count the small squares in a larger square.

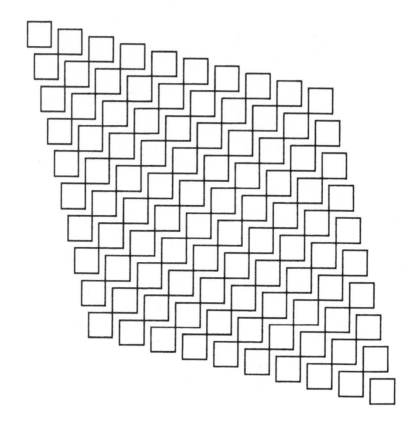

What happens if the squares are counted along the diagonals from south-west to north-east? In this picture the same square has been exploded in order to separate all those diagonals. This is the pattern of numbers that results:

$$1 + 2 + 3 + 4 + \ldots + 10 + 9 + 8 + 7 + 6 + 5 + 4 + 3 + 2 + 1$$

If we had started with a square of a different size, we would have found a different pattern of the same type. By imagining what would happen, we can predict with complete confidence, the sum of, for example,

$$1 + 2 + 3 + 4 + \ldots + 14 + 15 + 14 + 13 + \ldots + 3 + 2 + 1$$

We need make only one calculation: we find the middle number, which is 15, and multiply it by itself: $15 \times 15 = 225$

Ah! The power of mathematics!

Problems

1 Sum this series quickly by writing it a second time, underneath itself and *backwards*.

$$1 + 2 + 3 + 4 + 5 + 6 + 7 + 8 + 9 + 10$$

2 How can you deduce that $\frac{1}{3} + \frac{1}{9} + \frac{1}{27} + \frac{1}{81} + \ldots$ sums to one half, from the figure on page 71?

3 How can the diagram on page 68 be divided into *two* parts which will show that the sum of $\frac{1}{4} + \frac{1}{16} + \frac{1}{64} + \ldots$ is one third?

4 How can the series

$$1 + 2 + 3 + 4 + 5 + 4 + 3 + 2 + 1$$

be rearranged, or regrouped, to show that it has the same sum as: $1 + 3 + 5 + 7 + 9$, without finding the sum of either series?

5 Draw a picture to show the sum of this series: $1 + 8 + 16 + 24 + 32 + 40$

6 How can the diagram on page 68 be divided into a sequence of similar shapes to show that $\frac{3}{4} + \frac{3}{16} + \frac{3}{64} + \ldots$ sums to 1?

altitude

branch

island

ladder

field

queue

handle

lattice

net

tree

skeleton

saddle

family

face

foot

tandem

chain

pedal

pencil

sieve

envelope

ring

74

10

Metaphors

Why is old age the evening of life? The Greek philosopher Aristotle had a nicely mathematical explanation. He called it by the same Greek word we use today, metaphor, and he explained that a metaphor has the form of a proportion:

Old age **is** to life **as** evening **is** to the day.

Victorian children at school were expected to understand proportions in arithmetic presented like this:

12 : 8 :: 9 : 6

This was read: 12 is to 8 as 9 is to 6. Modern pupils might write the same fact like this: $\frac{12}{8} = \frac{9}{6}$

This can also be turned round to become: $\frac{12}{9} = \frac{8}{6}$ or 12 : 9 :: 8 : 6

So can Aristotle's metaphor: Old age is to the evening as a lifetime is to the day.

Aristotle praised metaphors and compared good metaphors to enigmas (and to jokes). To understand a metaphor is to solve a riddle. A metaphor creates new meaning. For this reason metaphors can be used to name what has otherwise no name. He offered as an example a famous enigma:

I saw a man who glued bronze with fire upon another.

There was no word in Aristotle's Greek to describe how a heated bronze cup, applied to an injury, would adhere by suction to the wound and draw out blood, so the word 'glued' was used, but how successfully! It not only suggests that the cup is firmly attached to the arm, but implies also that it can only be pulled away with difficulty.

Such metaphors, Aristotle continued, teach us by the new information they contain, provided they are not so obvious that there is nothing to work out or so difficult and obscure that the hearer misses the point entirely.

More than two thousand years after Aristotle, the Apache Indians used more than a dozen terms metaphorically to name the parts of an automobile, and no doubt to make the parts more easily learnt by children who saw a car for the first time. The front bumper was the 'daw' which otherwise meant 'chin and jaw'. The front wheel was the 'gun' which meant 'hand and arm'. The rear wheel was 'ke' or foot. A headlight was naturally 'inda' or eye. The hood was the 'chee' or nose.

They looked at the entrails of the car also. Electrical wiring was 'tsaws' or vein. The radiator hose was the 'chih' or intestine and the petrol tank was of course the 'pit' or stomach.

Mathematicians also use metaphors to name the nameless, to point to hidden connections, and to aid their memories by avoiding new names which mean nothing in themselves. The same metaphors teach newcomers by relating the mathematical ideas to their previous experience.

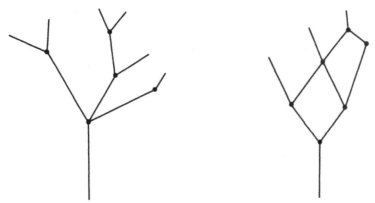

Open a book on graph theory and you will see diagrams like that on the left. Such a diagram is called, simply, a tree. Surprise!

The right-hand diagram looks less like a tree, because the branches are not separate but join and cross. This is not a tree. Mathematicians define a tree to be a graph with no circuits, with no loops, and therefore a tree whose branches do not join.

Mathematicians naturally talk about their trees. What could be more natural than to describe the branches as branches, or the point where the tree starts as its root – if it has a root. Some mathematical trees have no roots. Well, it takes all sorts . . .

What is a collection of trees called? Yes, you guessed it, mathematicians call a set of trees a forest. Must mathematicians stop there? Certainly not. Two papers in the prestigious *Philosophical Transactions of the Royal Society* a few years ago were titled, 'Periodic forests of stunted trees' and 'Periodic forests whose largest clearings are of size 3'.

The word 'tree' has always been a powerful metaphor. We talk of a genealogical tree of all our ancestors. Computers have ancestors also, the earlier computers to which they are related. A computer tree has been published showing hundreds of computers descended from the Mark 1 and the ENIAC. The computer tree has branches, like a family tree.

Mathematicians cannot always find everyday words to steal metaphorically for their new concepts, but they have not done too badly, as a check of a mathematical dictionary reveals:

aberration absolute affect altitude base box braid branch carrier chain character degenerate degree domain edge elastic envelope exhaustion face faithful family fibre field filter final fitting flat flow forgetful form foundation frame gap germ global gluing handle hereditary inseparable interior interval island jet join joint jump kernel killing ladder lattice layer layout leaf lift loading loop map member minor model natural neighbourhood net obstruction open operator passive pedal pencil peninsula pivot quality queue rank region regular replica residue ring root saddle separation sheaf shear shift sieve skeleton slackness tandem tolerance tooth trace trap tree trivial umbilical union universal vague vanishing variety web wedge weight wild

Most of the above mathematical metaphors are dying if not already dead. That is, the professionals who use them do not consciously think of their original meanings.

Other metaphors are used only when necessary, perhaps only by imaginative teachers and writers, and they never become standard terms. The word 'colour' is not in the above list, yet it is often used to clarify the expression of a problem, and maybe the problem's solution also.

Bruno Brooks on Radio 1 once proposed this problem to his listeners: 'If all even numbers are blue and all odd numbers are red, what colour is blue added to red?' What a masterly use of colour!

Donald Newman poses this problem in a book for undergraduates: 'The points of the plane are each coloured red, yellow or blue. Prove that there are two points of the same colour which are one unit apart.' If Newman had wanted to be totally conventional, he could have started: 'The points of the plane are divided into three disjoint sets . . .' How colourless!

The use of the word 'skeleton' to mean 'structure' is a metaphor. It is chosen to bridge the gap between the everyday understanding of the idea of structure, the bones that support the body, or the steel skeleton within the concrete pillars of a modern skyscraper, and the more abstract mathematical idea. Wallpaper patterns often illustrate skeleton and structure in both senses.

The very words 'isomorphic' and 'metaphor' can be interpreted as isomorphic, as metaphors of each other, provided that metaphor is interpreted in Aristotle's sense as a proportion.

×	+P	−P
+Q	+PQ	−PQ
−Q	−PQ	+PQ

+	even	odd
even	even	odd
odd	odd	even

When a mathematician says that these two structures are isomorphic, all that is meant is that there is a proportion between them. The relationship can be described by saying that, 'even is to positive numbers as odd is to negative as addition is to multiplication'. That it takes (at least) two proportions to describe the relationship illustrates the complexity that so easily appears in mathematics.

A similar complexity appears in proverbs such as, 'A rolling stone gathers no moss'. The listener must appreciate, in effect, that 'stone is to man as rolling is to constantly on the move as moss is to (whatever the speaker and the context implies)'.

It is typical of literature and everyday life that the interpretation of 'moss' is very vague without a context. Everyday language and literature use metaphors and other figures of speech, as Aristotle suggested, to create new meanings, to teach, to communicate (not always successfully). The underlying structure is only a means to an end.

Mathematicians are fascinated by the underlying structure itself, and they often get rid of the original contents in order to see the structure more clearly. Given a piece of wallpaper, a carpet or a length of curtain material, the mathematician looks past the flowers, the whorls, the decoration, to the underlying pattern.

It is no wonder that when the everyday clues have been stripped away mathematics often seems unrelated to life, but no wonder either that because these structures and hidden connections are found everywhere, the same abstract mathematics can later be filled with meaning in so many different ways, like building a sculpture on an armature.

Problems

1 'The product of two numbers is to the numbers as the area of a rectangle is to the lengths of its sides.' True or false?

2

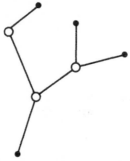

'The rule, "End points plus branch points equals number of edges plus 1" is to trees as Euler's relationship, "Vertices plus regions equals edges plus 2" is to plane maps.' True or false?

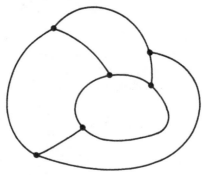

3 Number pairs, such as (1, 5) and (9, 3) can be added by simply adding their parts separately, like this: (1, 5) + (9, 3) = (1+9, 5+3) = (10, 8) Number pairs can also be pictured on a graph:

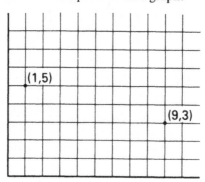

What is the geometrical equivalent on such a graph of addition by adding the parts separately?

4 The formula for the nth triangular number is $\frac{1}{2}n(n+1)$. The formula for the nth square number is, of course, n^2 and the formula for the nth pentagonal number is $\frac{1}{2}(3n^2 - n)$.

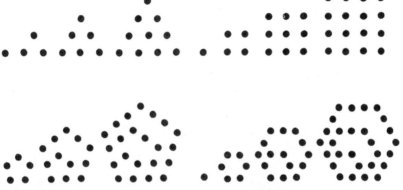

Complete the proportion: '$\frac{1}{2}n(n+1)$ is to triangular as n^2 is to square as $\frac{1}{2}(3n^2 - n)$ is to pentagonal as . . . is to hexagonal.'

5 A rooted tree is, not surprisingly, a tree with one end point designated as the root, as in these trees with 1, 2 and 3 edges respectively.

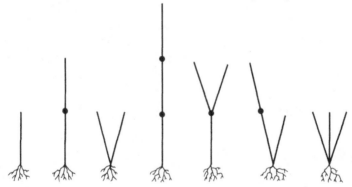

How does the number of different rooted trees with 1, 2, 3, 4, . . . edges compare with the number of arrangements of soap bubbles, as in this illustration?

6 Complete this proportion: 'The average of two numbers is to the numbers as the . . . is to the positions of two numbers on a ruler or number line.'

A cross section of a nautilus shell.

One problem, many solutions

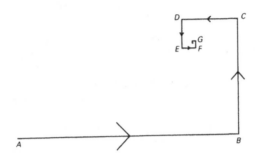

We can think of this spiral as the path of a bug which turns left more and more frequently. Intuitively it seems that the bug is going somewhere, it is moving towards some centre which we shall naturally call the limit, or limiting point, of the spiral. Following the example of Chapter 9 we look for a point which it approaches more and more closely, to which indeed we can get as close as we like.

It turns out that this limiting point of the spiral can be found in many ways, depending on how we 'see' the spiral. For example, look at the first two movements together. In fact, replace them by one straight line movement.

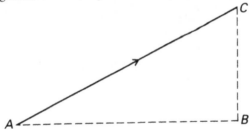

Now look at the next two movements, and replace them by a straight line also.

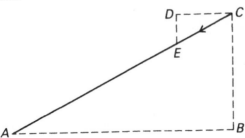

83

By looking at the movements two at a time, the total movement is much simplified. It is simply to and fro in the same diagonal direction. What is the sum total of this diagonal movement? Call the first movement from A to C, 1 unit. The movement backwards is one quarter of this, and the next movement forwards is one quarter of that, and so on . . .

The total movement along the diagonal, counting AC as one unit, is:

$$1 - \tfrac{1}{4} + \tfrac{1}{16} - \tfrac{1}{64} + \ldots$$

The sum of this infinite series can be discovered by the methods of Chapter 9, and turns out to be $\tfrac{4}{5}$. So the limiting point of this spiral is apparently $\tfrac{4}{5}$ of the way from A to C.

Suppose that we are not satisfied to find the limiting point by such a mixture of geometry and arithmetic, but wish to find it by geometry alone. This can be done as well. It is already clear that the limiting point lies somewhere on line AC. Suppose that the bug started its journey at B, not A. The limiting point has not changed at all. If we consider the movements from B in pairs, first moving directly from B to D, then we shall conclude by exactly the same argument that the limiting point lies on the line BD.

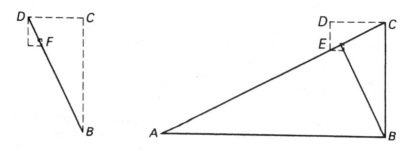

Or suppose that the spiral started at C. The limit point is unmoved, and we see that it must lie somewhere on CE, which was one of the lines in our first construction. Therefore we can conclude that the lines AC, BD, CE, and so on, all pass through the same point, which must be the limit point. Making an actual drawing and joining these points confirms experimentally that this is so.

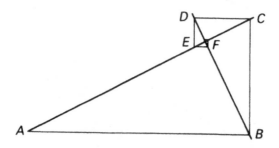

There is yet another way to think of the spiral geometrically. Imagine that the limit point is known. Mark it and join it to A, B and C.

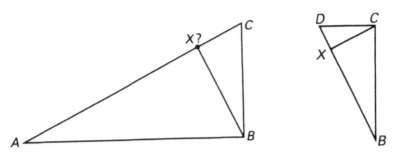

Now think of the right-hand diagram, as the left-hand diagram after a compass point has been stuck into the limit point, and the whole figure has been rotated through a right-angle and reduced to half its size. Naturally, AB has now become BC. BC has turned into CD.

What about the individual triangles? Each triangle is one half the size of the previous triangle, in length, and is at right-angles to it. So the limit point can be found by searching for a point which makes triangle AXB similar to triangle BXC. This final problem can be solved by geometry alone. (In fact, since ABC is a right-angle, X is the foot of the perpendicular from B to AC.)

Geometry, however, may not appeal to you. Perhaps you would prefer to solve the problem by arithmetic instead? This is also possible. We go back to the original diagram, and think of the horizontal movements and the vertical movements as slightly separated.

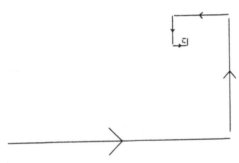

The horizontal movements by themselves, are 1 to the right, $\frac{1}{4}$ back to the left, $\frac{1}{16}$ to the right, and so on.

$$1 - \tfrac{1}{4} + \tfrac{1}{16} - \tfrac{1}{64} + \ldots$$

This is a series that has already been summed. Its limit is $\frac{4}{5}$.

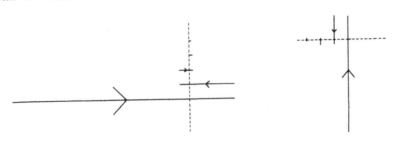

What about the vertical movements? These are
$$\frac{1}{2} - \frac{1}{8} + \frac{1}{32} - \frac{1}{128} + \ldots$$
Every term in this series is exactly one half of the matching term in the previous series, and so the limit of this series will be only $\frac{2}{5}$.

Put the two series together again and the limit point is found $\frac{4}{5}$ to the right and $\frac{2}{5}$ up.

Many other spirals can be studied in the same way. The frontispiece to this chapter shows a nautilus shell, whose spirals are related to the Fibonacci numbers. A good approximation to the spiral of a nautilus can be drawn like this.

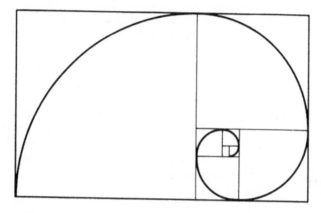

The original rectangle is almost exactly 1 by 1.618 units. This rectangle is special, because if a square is cut off one end, the rectangle that remains is the same shape as the original one. So another square can be cut off . . . and so on.

Now draw quarter circles in each of the squares with a compass. This spiral is imperfect, but it is a good approximation to the nautilus spiral. Where is its centre? Where is the limit point?

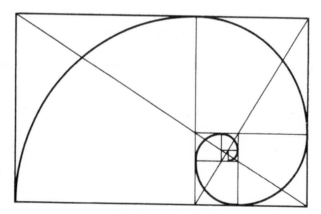

Join the vertices of the square spiral like this, and they all go through the centre of the spiral.

Problems

The following problems all concern this question:

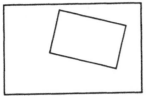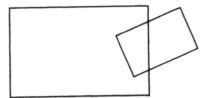

'Two maps which are identical except that they are to different scales, are placed one on top of the other, as in these diagrams. Must it be true that there is one point on the large map which is directly over its corresponding point on the small map?

Which of the following arguments is correct?

1 Cover each of the maps by a square grid, making the size of the unit squares proportional to the scale of each map. Number the lines of the grids identically so that each point on each map has two grid-reference numbers. Then you can find the point which has the same grid-reference numbers on each map.

2 In the one-dimensional case, by marking each line as if it were a ruler, it is obvious that there must be a point with the same marking, because the right-hand end of the short line has a higher number than its matching point on the long line, while its left-hand end has a smaller number than its matching point.

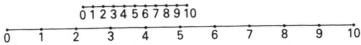

Since the numbers on both lines change continuously it is obvious that somewhere in between they will be equal.

The same argument applies to the two-dimensional case.

3 There need not be a matching pair of points because the two maps might not overlap sufficiently, as in this diagram.

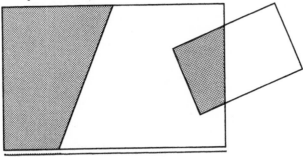

Any matching point must be in the shaded area of the small map, which matches the shaded area of the large map, and these do not overlap at all.

4

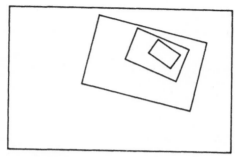

Imagine that a third map, to a proportionately smaller scale, is placed on the small map, exactly as the small map is placed on the large map . . . and this process is repeated. Then the point where the original two maps agree will also be a point where all the smaller maps also agree, and since they are tending to a point as they get smaller and smaller, that point is the limiting point.

5

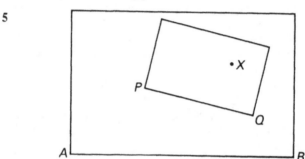

The common point, if it exists at all, will be in same relationship to the corners of the small map as it will to the same corners of the large map. So the problem is simply to construct the point X, such that triangle AXB is similar to triangle PXQ. X is then the required point.

6 The common point, if it exists, will be in the same relationship to the line AP considered as a part of the large map, as it will be to the line PF which is drawn to match AP exactly on the small map.

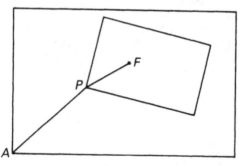

But the line segments AP and PF are the start of a spiral which tends to a limit point, which will be the common point of the two maps. So it is only necessary to find the limit point of this spiral.

Pictures into numbers

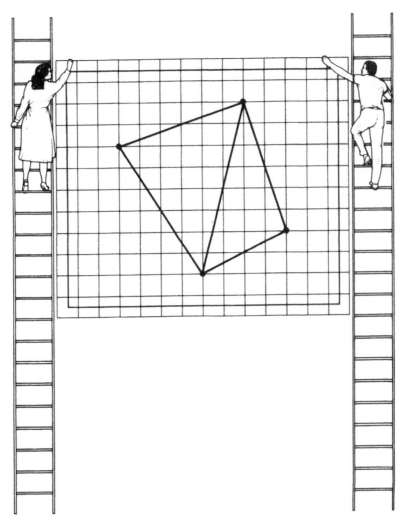

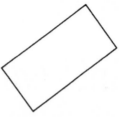

This rectangle was drawn with a set square to make sure that the corners were right-angles. This one feature is sufficient to guarantee that it is indeed a rectangle and not some other shape.

However, rectangles have many other properties. For example, their opposite sides are always equal in length and parallel, their two diagonals are also equal in length, and the point where the diagonals meet is the same distance from each corner.

These facts can be proved by the kind of geometry found in Euclid's two-thousand-year-old textbook called *The Elements*. Amazing to relate, however, they can also be proved by first translating the rectangle into a pattern of numbers, or rather pairs of numbers. Indeed, it turns out that any statement at all about distances or angles or areas of a geometrical figure can be translated into statements about numbers. There is a perfect isomorphism, or metaphor, between traditional geometry and the properties of pairs of numbers, provided the 'dictionary' needed to make the translation is understood.

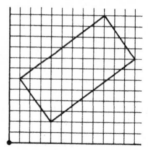

The first step is to place a square grid over the geometrical figure which we wish to study, much as a grid of lines of latitude and longitude is placed over a geographical map. Just as this grid gives every point of the map a map-reference, so every point of the rectangle can now be described by a pair of numbers. Simply start at the dot, which is called the origin, and count right, and then up, to reach the point that is to be described.

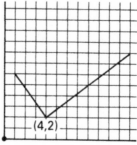

(4,2)

The lower left corner of the rectangle is 4-right-and-2-up from the origin, and is described as the point (4, 2) The other three corners are described as in the next diagram.

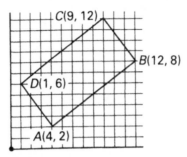

Now we can start to fill in the dictionary which translates each geometrical property into a property of numbers, and it is immediately apparent just how powerful this method can be. In the geometrical diagram the sides AB and DC are equal in length and also parallel to each other.

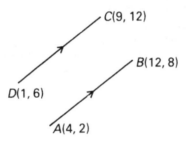

Both these facts together can be described by saying that from $A(4, 2)$ to $B(12, 8)$ is 8-right-and-6-up, and from D to C is also 8-right-and-6-up.

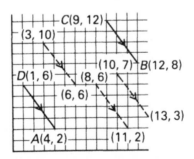

DA and CB are also parallel to each other, and equal in length. The translation of these properties is, 'From D to A is 3-right-and-4-down, and from C and B is also 3-right-and-4-down.' In order to emphasise how general this translation is, the diagram shows several pairs of points with this property, and sure enough the lines joining them are all parallel and equal in length.

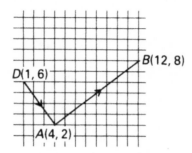

What about the right-angles at the corners of the rectangle? These are also easily interpreted using the same language, though the result is a little more complicated. *DA* is at right-angles to *AB*. From *D* to *A* is 3-right-and-4-down, and from *A* to *B* is 6-up-and-8-right. 'Right-and-down' has been exchanged for 'up-and-right' and the numbers have been doubled.

Exchanging right-and-up for down-and-right while keeping the numbers in proportion always creates a right-angle.

How can the middle point of a line be translated into numbers? This is very simple indeed. What number pair is half way between (4, 2) and (9, 12)? Half way between 4 and 9 is $6\frac{1}{2}$, and half way between 2 and 12 is 7. The middle pair or point, half way between *A* and *C*, is $(6\frac{1}{2}, 7)$.

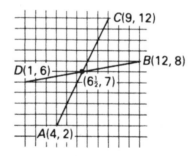

Where is the point half way between *B*(12, 8) and *D*(1, 6)? Half way between 1 and 12 is $6\frac{1}{2}$, and half way between 6 and 8 is 7, so this middle point is also $(6\frac{1}{2}, 7)$ It is no surprise that the two middle points are the same!

So much for parallel lines, lines at right-angles, and middle points, but what about distances? Are not distances fundamental to geometry? Indeed they are, and it is not surprising, bearing in mind that we are using a grid of lines at right-angles, that they are translated into the language of number pairs by using the oldest, most traditional of all geometrical theorems, which was known to the Babylonians long before it was discovered by Pythagoras.

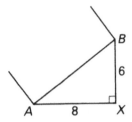

Here is the bottom edge of the rectangle, viewed as the hypotenuse or longest side of a right-angled triangle. By Pythagoras' theorem,

$$AB^2 = AX^2 + XB^2 \text{ or } AB^2 = 8^2 + 6^2 = 100,$$

and so $AB = 10$ units.

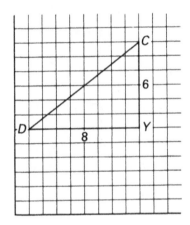

Is DC the same length? In this right-angled triangle

$$DC^2 = DY^2 + YC^2 = 8^2 + 6^2 = 100,$$

so DC also equals 10 units. Now we can see why every movement 8-right-and-6-up produces a line of the same length, wherever we start. Every such movement is the longest side of the same sized right-angled triangle. How elegantly all the pieces fit together!

We will fit one more piece of the jigsaw puzzle into place here. How can the area of the rectangle be calculated? The longest side is 10 units and the shorter side turns out to be 5 units, so the area is $10 \times 5 = 50$. However, we do not need to go to such trouble, and in particular there is no need to calculate the lengths of the sides. It is quite sufficient to look only at the movements 8-right-and-6-up and 3-along-and-4-down which describe the sides, and to calculate

$$8 \times 4 + 6 \times 3 = 32 + 18 = 50$$

Magic!

Problems

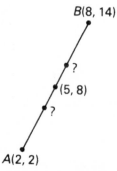

B(8, 14)

?

(5, 8)

?

A(2, 2)

1 The middle point of this line is (5, 8), half way between $A(2, 2)$ and $B(8, 14)$. How can the point which is one third of the way from A to B be calculated? Where is the point one third of the way from B to A?

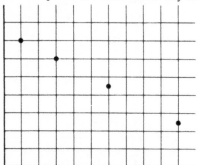

2 Some problems which can be solved instantly in a diagram are much harder to solve in numbers. It is immediately obvious that these points lie on a straight line.

Is it true that the points (6, 1), (16, 7) and (21, 10) lie on a straight line? How can this be tested without drawing?

3 These five points lie on a straight line, with one exception:

(1, 1) (4, 3) (7, 5) (10, 8) (13, 9)

Without drawing a diagram, which is the odd one out?

4 It is easy to 'add up' three number-pairs to find their total, like this:

(5,   2)
(9,   3)
(7,   7)
(21, 12)

It is then very easy to 'divide' the total by 3 in order to find the 'average' of the three original points: (21, 12) divided by 3 is (7, 4). What is the geometrical meaning of this arithmetical calculation?

5 Is it possible to find the 'average' of three points by finding first the 'average' of two of them? For example, if one point is (3, 8) and the 'average' of the other two is (6, 5), can the 'average' of all three points be calculated? What is the geometrical meaning of this calculation?

6 Is it true that the points (9, 8), (6, 11), (1, 12) and (10, 4) all lie on a circle whose centre is the point (2, 4)?

13
A point of balance

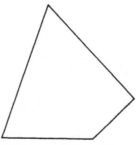

Here is a quadrilateral, a simple geometrical object. If I doodle, as mathematicians sometimes do, and join the mid-points of two opposite sides I can mark X as the 'centre' of the quadrilateral.

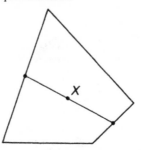

It cannot be denied however that this figure does not look very symmetrical. If I continue to doodle by joining the mid-points of the other pair of sides, I shall probably get another, different 'centre'.

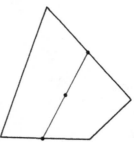

Let us do this little experiment. Surprisingly, both points coincide. It seems as if the quadrilateral **does** have a centre, which I can construct in either of two ways. How can this be? How can I **see** that this must be so?

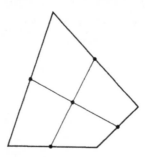

One approach is to exploit a parallelism, a metaphor or isomorphism, between this piece of geometry and a bit of physics. To be more precise, a bit of mechanics.

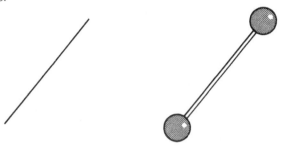

On the left is a line with its mid-point. On the right, two identical 1-kg weights are joined by a thin rod which I shall imagine weighs nothing to speak of. It does however have a mid-point, at which I could balance the dumb-bell on the point of a pin.

The weight pressing down on the pinhead is equal to 2 kg, since we are ignoring the negligible weight of the rod. Indeed, as far as anything the girl can feel, it might just as well be a single 2-kg weight.

Now let us translate the entire quadrilateral in the same way. The four corners become four identical weights, joined by four light rods. On second thoughts, let them be joined by a thin lightweight sheet, because then I can more easily imagine balancing them on a pinpoint at some point of the sheet.

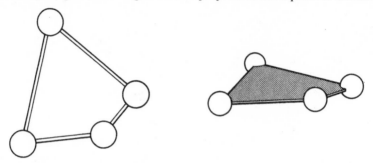

Where will the balance point be? I can imagine trying different points, and adjusting the quadrilateral until I find the point where, as near as makes no difference, the weights are supported without tipping in any direction. then I can take my hand away.

I am quite certain from my experience of balancing objects that there will only be **one** point where the whole contraption balances. But where will it be? This is where I can 'calculate' the point, by thinking of the dumb-bells.

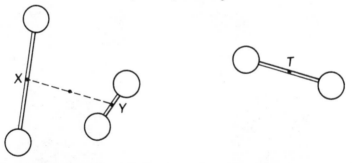

I imagine the left-hand dumb-bell, which might just as well be one 2-kg weight at X. I imagine the right-hand pair, which might just as well be a 2-kg weight at Y. Now we have another dumb-bell, which, however, is twice as heavy as the original one. Where will it balance? Naturally it will balance at its mid-point, where it will feel like a single 4-kg weight.

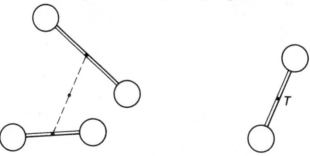

How about the second way of matching the weights? I could also imagine the top and bottom pairs replaced by single weights making another double-weight dumb-bell, which would balance at a point T.

What is the significance of this second solution? We already know that there are apparently two ways of finding the centre. What difference does the metaphor of the dumb-bells make? It makes a very big difference, because we have agreed, I hope, that the real weights, the dumb-bells at the corners of the weightless sheet, can only balance at ONE point. It is inconceivable that they could balance at two different points. That would contradict all our experience and intuitions of the way real weights behave. Therefore, precisely because we have been thinking of the real weights of the real dumb-bells, we can be certain that the two points T are indeed the same point, the unique point where the whole contraption balances!

There is a bonus waiting for us here. The original quadrilateral was drawn flat on the paper. There was no suggestion that it might be a three-dimensional quadrilateral. However, we have nowhere exploited the fact that it was flat. The arguments we have used would all have worked as well if the quadrilateral had been in three dimensions, as in this figure.

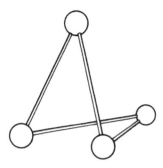

In imagination (how invaluable imagination is to the mathematician!) or by constructing an actual physical model, we can find the centre exactly as before. In how many ways can we do this? The answer is in not two but three different ways.

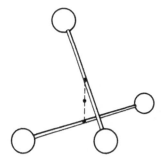

The third way of choosing the pair of dumb-bells corresponds to the choice of the diagonals of the original quadrilateral, a possibility which we previously ignored.

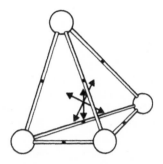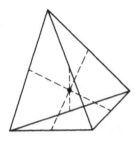

Now let us put the three ideas together. The complete figure is four weights joined now by six light rods. It is a metaphor for a triangular pyramid, or irregular tetrahedron. We can now translate our conclusion and state a theorem about any tetrahedron. Our impressive conclusion says that: 'The three lines which join the mid-points of opposite sides of a tetrahedron all bisect each other at a single point.'

Problems

1

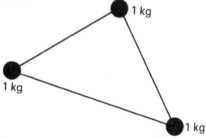

5 kg

3 kg

Assume that two equal weights are equivalent to their total weight at a point half way between them, where they will balance. Where would you expect these 3-kg and 5-kg weights to balance? The bar that holds them weighs nothing.

2

1 kg

2 kg

Making the same assumption as in problem 1, where will these 2-kg and 1-kg weights balance?

3

1 kg

1 kg

1 kg

Given the solution to problem 2, where will these 3 one-kilogram weights at the corners of a triangle balance?

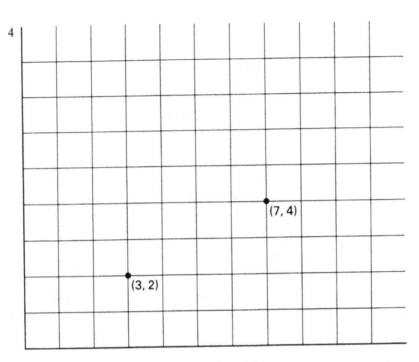

Two points, (3, 2) and (7, 4) have been marked on this graph. If equal weights are placed at each point, where will the weights balance?

Where would equal weights at (3, 2), (7, 4) and (2, 6) balance?

5

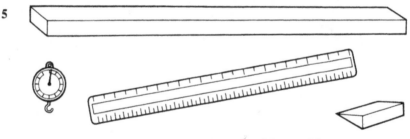

This is a long heavy uniform beam, which you wish to weigh. Unfortunately you only possess a small spring balance, capable of weighing a few kilograms, plus a ruler and a strong knife edge. How do you weigh the whole beam? (You are not allowed to cut it into pieces!)

6

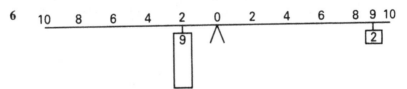

This balance rests on the knife edge at the point marked 0. Will the 9-kg and 2-kg weights balance each other?

Multiple meanings

How impoverished language would be if it lacked ambiguity and double meanings! How flat! How unfunny! Fortunately all languages are loaded with ambiguity, and in particular with figures of speech. Indeed, many linguists believe that all language is essentially metaphorical, and that a major source of the growth and development of language is the extension of meaning by metaphor.

For scientists this is almost a paradox. They like to be able to say precisely what they know, yet their new ideas often arise by interpreting what they know in a new way. It might be thought that mathematical language must surely be extremely precise. If mathematicians do not know what they are talking about, who does? In one sense, this is true. Any mathematician should be able to explain what he or she means by a particular statement in mathematical language, and this explanation should be clear and lucid. At the same time, it is also true that the same statement will be open to a variety of interpretations.

There is no contradiction here. Neither is this a weakness of mathematics. Quite the opposite! It is one of the greatest strengths of mathematics. If a mathematical statement could only ever be interpreted in one way, how many statements you would need!

How little information each would carry! Because they can be interpreted in so many ways, they are more informative and more useful and more powerful.

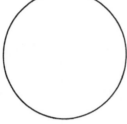

This is a common or garden circle, with radius 1. Its area as every girl and boy knows is given by the formula $A = \pi r^2$. What does this mean? One very easy and practical meaning is that to calculate the area of the circle, you simply multiply π by the value of r^2. Since π is approximately 3.14, and the square of the radius is 1, the area of the circle is approximately $3.14 \times 1 = 3.14$ square units.

This interpretation as a formula, an algorithm which tells you the steps that you must perform to make a correct calculation, is very useful, but it says nothing about why the formula works. This figure can be more illuminating.

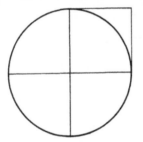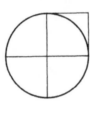

On the left is a circle and two diameters, and the square which fills one quadrant. The square has sides of 1 unit, so its area is also 1 square unit. The circle is of course bigger. How much bigger?

Look at the second circle, which also has a square in one quadrant. How does THAT circle compare with its square? Will the comparison not be the same in each case? If the area of the circle in the first figure is about 3 times the area of its square, will not the smaller circle be about 3 times in area the size of its smaller square? Indeed it will. Although the absolute areas change, their ratio does not. We can write this as a proportion.

area of large circle : area of large square = area of small circle : area of small square

We supposed that the area of each circle was about 3 times the area of its matching square. In fact the ratio is a little more than 3. It is the number called π or approximately 3.14159. What does this point of view tell us about a general circle and its area? The area of the square in its quadrant is r^2, where r is the radius of the circle. The formula tells us that the area of the circle is π times as big, because **all circles** are in area π times the area of a square in one quadrant. So the area of the circle is equal to πr^2.

This is one explanation of the formula, which says more than any merely algorithmic interpretation. However, it is not the only interpretation.

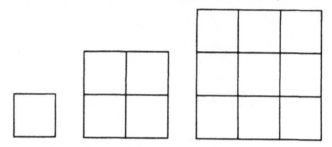

Here are three squares. The middle square is twice as tall and twice as wide as the small square. It area is $2 \times 2 = 4$ times as great, as shown by the small squares into which it is divided. The large square is three times as tall and three times as wide as the small square, and its area is $3 \times 3 = 9$ times as great. The large square is also $1\frac{1}{2}$ times as wide as the middle square and its area is $1\frac{1}{2} \times 1\frac{1}{2} = 2\frac{1}{4}$ times as great. As a check, notice that $9 = 2\frac{1}{4} \times 4$.

It is a general rule that when a figure is enlarged by a factor, say 2.6, (preserving its shape) then its area increases by the same factor squared: $2.6^2 = 6.76$. The same principle is illustrated by problem 5 of the first chapter.

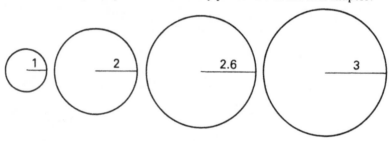

Circles are no exception. A circle of radius 2 is $2^2 = 4$ times the area of a circle of radius 1. The circle of radius 3 will be 9 times the area. The circle radius 2.6 is $2.6^2 = 6.76$ times the area of the unit circle. In contrast to the previous interpretation, there are now no squares in sight, only circles. There is only one difficulty. We need to know the area of a unit circle. What is it? Well, the area of a unit circle is 3.14159 . . . and so the area of a circle of radius 2 is 4 times 3.14159 . . . and so on.

In the first interpretation π was a ratio between a circle and a square. In the second interpretation it is an area itself, the area of the unit circle. Is not one interpretation 'correct' and 'right' and 'true'? Is not one of them better than the other? Certainly not! They are both equally good, and merely different ways of looking at the same formula.

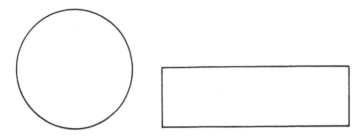

There is yet a third way to look at the formula. πr^2 can be written as $r \times \pi r$, which could be the area of a rectangle r units in one direction, and πr in the other. Does this make sense? Can it make sense, from a suitable point of view? Indeed it can. The brilliant Greek mathematician Archimedes first realised that if a circle were divided into lots of small slices, like a cake being divided among many people, then the slices could be rearranged to make an approximation to a rectangle.

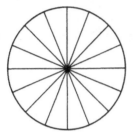
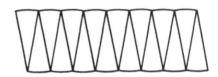

The more slices that are cut, the closer each slice is to a triangle, and the nearer the rearrangement is to a rectangle. How tall is the rectangle? Its height is r, the radius of the circle. How long is it? One half of the circumference of the circle, which is therefore one half of $2\pi \times r$, according to another well-known formula. So the rectangle has area $r \times \pi r = \pi r^2$, as expected.

From each point of view something has been left unexplained. In the first interpretation, it was **why** the circle is approximately 3.14159 . . . times as large as the square drawn in one quadrant. In the second interpretation, it was **why** the area of a unit circle is approximately 3.14159 . . . square units. In this third interpretation it is **why** the distance round a circle is $2\pi r$ or approximately 6.28318 . . . Why? That is another story!

Problems

1 To which of these questions is $\frac{3}{5}$ the correct answer?
   (a) What is $\frac{1}{5} + \frac{1}{5} + \frac{1}{5}$?
   (b) What is 3 divided into 5 equal parts?
   (c) How many times does 5 divide into 3?
   (d) What is a fifth of 3?

2 Which of the following interpretations of this equation is correct?

$$5x - 6 = 2x$$

(a) $5x$ is the sum of 6 and $2x$.
(b) From $5x$ down to $2x$ is 6.
(c) The difference between $5x$ and 6 is $2x$.

3

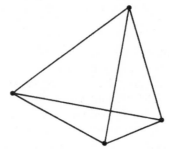

Is this:
(a) a picture of a tetrahedron?
(b) 4 points and the 6 lines joining them in pairs?
(c) two triangles with a common edge with their free vertices joined?

4

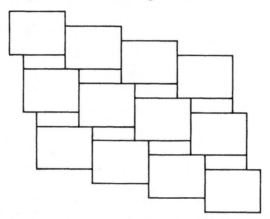

Is this:
(a) a tessellation composed of two sizes of rectangles
(b) a tessellation of identical rectangles in which every row, horizontally and vertically, as it were, has been slipped, to leave a pattern of rectangular holes?

5 Does the description, 'A polygon all of whose sides are equal and all of whose angles are right-angles', define
(a) a square?
(b) a rectangle?
(c) neither of these?

6 Which of the following is a sound interpretation of negative numbers?
(a) angles measured clockwise are +ve and angles measured anti-clockwise are −ve.
(b) assets are +ve and debts are −ve
(c) temperature above zero degrees are +ve and temperatures below zero are −ve.

Matching problems

Everyone is familiar with machines and devices that the makers guarantee will work every time, if you follow the instructions correctly. A key in a lock is a very simple example, so simple that instructions are hardly necessary. Slightly more complicated are keys that must be turned two or three times to release the bolt. Door answerphones are rather more complex, and washing machines are another level up.

Library computer terminals, which allow you to discover whether a book is on the shelves, which books you have out and which are overdue, are more complicated still. They are easy to use only because the screen prompts the user by listing the next available moves. Knitting patterns can seem mindboggling to the uninitiated, although the instructions are individually clear and unambiguous and only have to be followed in sequence.

Many mathematical problems can be solved in the same way. The following equations could have originated in a problem in physics or chemistry, in economics or the social sciences, or from a problem in pure mathematics. Whatever their sources, they are easily solved by the same well-known routine processes, called algorithms.

$$10.749u - 1.44v = 2$$
$$6.05u + 39.81v = 1$$

$$2x + 3y - 5z = 7$$
$$4x - y + 2z = 10$$
$$7x + 4y + 6z = 9$$

$$12a \quad\quad -c \quad\quad = 1.43$$
$$8a - 13b \quad\quad + 8d = 7.50$$
$$8b \quad - 2d = 7.91$$
$$4b - 6c + 3d = 1.59$$

To a mathematician all these sets of equations are essentially 'the same'. The numbers in each are different as are the letters which appear but the methods of solution of each are effectively identical and a computer soon churns out the answers, if they exist.

To realise that these equations are all of one type is not difficult. It is only necessary to spot a simple pattern, much like spotting the pattern that makes both these verses limericks.

A spider's more legs than a flea,   I walked by the river all day,
This knowledge I give to you free.   Throwing stones at the fishes at play,
If you find it of use,   Five fisherman offered,
And not merely abstruse,   To drown me, I proffered,
Remember you got it from me.   Five thank-yous and strolled on my way.

Spotting that one poem is a parody of another can be more subtle, but not in the following example. Robert Southey was a serious poet, but Lewis Carroll was more struck by Southey's scansion than by his sentiment. Southey started and ended thus,

'You are old, Father William,' the young man cried,
    'The few locks which are left you are grey;
You are hale, Father William, a hearty old man,
    Now tell me the reason, I pray.'

'I am cheerful, young man,' Father William replied,
    'Let the cause thy attention engage;
In the days of my youth I remembered my God!
And He hath not forgotten my age.'

Lewis Carroll responded with:

'You are old, Father William,' the young man said,
    'And your hair has become very white;
And yet you incessantly stand on your head –
    Do you think, at your age, it is right?'

'In my youth,' Father William replied to his son,
    'I feared it might injure the brain;
But, now that I'm perfectly sure I have none,
    Why, I do it again and again.'

Many essentially identical problems in mathematics are very easy to spot because of just such simple patterns. The most common problems in applications of mathematics are of this kind. The same standard types turn up again and again and the same algorithms are used to solve them. New problems are continually being discovered for which new algorithms are worked out. This is one source of the power of mathematics which turns so many initially difficult problems into routine exercises.

However, not all problems which are essentially identical appear to be so at first sight. Their common structure may be very well hidden. For example:

Problem 1: How many triangles are there in this figure?

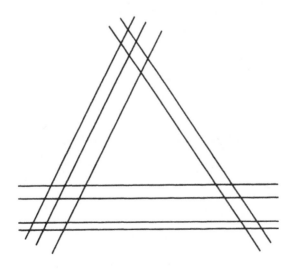

Problem 2: In how many ways can a casting director choose a mother, a father, and one child, from 2 actresses, 3 actors and 4 children?

Is the connection between these problems not obvious? Then let us start by solving the casting director's problem by a crude but effective method – writing down all the possibilities.

In order to not miss any possibilities, it is wise to list them according to a simple system, so let us first give the characters names. Let the actresses be called Anne and Barbara, the actors Charles, David and Edward, and the children, Frank, George, Helen and Jane. Using their initials only, the casting director's possible choices can be listed in this order:

ACF	ACG	ACH	ACJ	BCF	BCG	BCH	BCJ
ADF	ADG	ADH	ADJ	BDF	BDG	BDH	BDJ
AEF	AEG	AEH	AEJ	BEF	BEG	BEH	BEJ

The system behind this listing is quite simple. If Ann is the mother, then each of the men, C, D and E, can be matched with any of the children, F, G, H and J. If Barbara is the mother, the same pattern is repeated.

The casting director has 24 choices, which could have been calculated as 2 choices of mother multiplied by 3 choices of father multiplied by 4 choices of child, making a total of $2 \times 3 \times 4$ possible choices in all.

Counting the triangles is not quite so simple. However, a little insight will help. Because the lines are parallel in three sets, we may notice that any triangle MUST include one of the bottom four lines. It must also include one of the two right-hand lines, and one of the three lines on the left.

They are now easier to count, but there is no way to write down each triangle as we count it, unless we label the lines. So let us give them names.

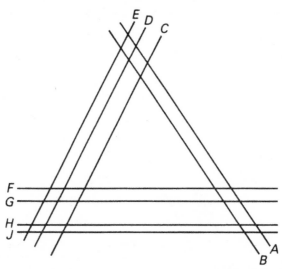

It is no coincidence that the letters chosen to label the lines are the initials of the names in the casting director's problem. The possible combinations happen to be identical also:

ACF	ACG	ACH	ACJ	BCF	BCG	BCH	BCJ
ADF	ADG	ADH	ADJ	BDF	BDG	BDH	BDJ
AEF	AEG	AEH	AEJ	BEF	BEG	BEH	BEJ

It is now clear that both problems are essentially the same, despite their different 'dressings'!

These problems are also isomorphic, though in a slightly more subtle way.

Problem 1: How many numbers are equal to the sum of three of their factors that are all different?

Problem 2: In how many ways can one be expressed as the sum of three different reciprocals?

Thinking of possible answers to the first problem, it is easy to spot numbers such as 12 and 24 and 36 which have many factors each, and can certainly be expressed as the sum of only three of them:

$$12 = 6 + 4 + 2 \qquad 24 = 12 + 8 + 4 \qquad 36 = 18 + 12 + 6$$

In contrast there are very few ways to express 1 as the sum of three reciprocals, because most reciprocals are so small.

$$1 = \tfrac{1}{3} + \tfrac{1}{3} + \tfrac{1}{3}$$

is disallowed because the reciprocals must be different but it does emphasise that not all the reciprocals can be less than or equal to $\tfrac{1}{3}$, and that therefore one of them must be greater. It follows that one of them must be $\tfrac{1}{2}$ and since

$$1 = \tfrac{1}{2} + \tfrac{1}{4} + \tfrac{1}{4}$$

is also forbidden the only possible sum is

$$1 = \tfrac{1}{2} + \tfrac{1}{3} + \tfrac{1}{6}$$

How does this problem of the reciprocals relate to the sum-of-factors problem? In particular, how does this unique solution square with the many solutions of the first problem?

The answers to both questions can be seen by dividing the numbers in the solutions to the 'factors' problem, like this:

$$\tfrac{12}{12} = \tfrac{6}{12} + \tfrac{4}{12} + \tfrac{2}{12} \qquad \tfrac{24}{24} = \tfrac{12}{24} + \tfrac{8}{24} + \tfrac{4}{24} \qquad \tfrac{36}{36} = \tfrac{18}{36} + \tfrac{12}{36} + \tfrac{6}{36}$$

When all these fractions are reduced to their lowest terms the equation which appears is always '$1 = \tfrac{1}{2} + \tfrac{1}{3} + \tfrac{1}{6}$'. Turning this idea round, we can obtain a solution to the first problem by multiplying the numbers, $1, \tfrac{1}{2}, \tfrac{1}{3}$ and $\tfrac{1}{6}$ by any number which is divisible by 2, 3 and 6, because this will leave a sum with only whole numbers and no fractions.

What are the numbers which are divisible by 2, 3 and 6? Why, they are 6 itself and all multiples of 6. Multiplying by 6 gives this solution to the first problem: $6 = 3 + 2 + 1$. Multiplying by 12, 24 and 36 gives the solutions listed already. In between and extending to infinity are the other possible multiples, 18, 30, 42, 48, 54 . . . An infinite sequence of numbers all effectively hiding the fact that $1 = \tfrac{1}{2} + \tfrac{1}{3} + \tfrac{1}{6}$.

Problems

1

This puzzle was first presented by Guarini in 1512, according to H.E. Dudeney who discusses it in his *Amusements in Mathematics*. The problem is to make the two white knights exchange places with the two black knights in as few moves as possible. All the knights moving as in chess, and any number of successive knight's jumps by the same piece counts as one move.

Your task is slightly different, to discover a new way of looking at the puzzle which makes it even easier to solve.

2 These six problems are actually isomorphic in pairs. In other words, solve three of them and you will have solved all six, provided you understand which problems are equivalent to each other. How do they pair up?

(a)

The four binary numbers 00, 01, 10 and 11 can be arranged in this circular sequence, in which each pair of adjacent numbers has one pair of matching digits the same and one pair different.

In how many ways can these eight binary numbers, 000, 001, 010, 011, 100, 101, 110, and 111 be arranged in a similar circular sequence, so that every pair of adjacent numbers has the digits in two places in common? For example, 010 and 011 could be adjacent, because they share the same first two places, but 001 and 100 must be separated.

(b) Solve this quadratic equation: $t^2 + 10 = 10t$
(c) What is the maximum area of a rectangle if the sum of its four sides is 20?
(d) I think of two numbers. Their sum is 10 and their product is 10 also. What are the numbers I thought of?
(e) How many routes are there from one corner of a cube, visiting every other corner exactly once, and returning to your starting point?
(f) The sum of two numbers is 10. How big can their product be?

The game of algebra

Algebras can be thought of as languages for talking about mathematical objects and their properties. There is an algebra of ordinary numbers which is the first algebra taught in schools. There is an algebra for forces in mechanics which, like velocities, need to be described by their size and direction. There is an algebra for the way a cube can be taken out of a cubical box and put back again. Surprisingly, this algebra is not a trivial curiosity but turns up in the thinnest of disguises in many other situations.

There are even algebras for talking about other algebras. Unlike ordinary language, however, algebras have the special property that from one algebraic statement you can generate many more by sequences of very simple operations. If the statement you start from is true, then so will all the statements inferred from it. This is a powerful means of discovering new truths, especially when combined with an emphasis on what algebraic statements are saying in ordinary language.

Many algebraic sentences can be 'seen' in different ways. 'Insight' and 'vision' and 'looking ahead', 'seeing what is happening' are just as important in algebra as they are in geometry, or in the algebra-with-geometry which is what mathematicians use much of their time. The verb 'to see' may be used by analogy only, as a metaphor, but the activity it describes is nevertheless real although there need be and often are no pictures or diagrams in algebra.

Suppose that we wish, either for a serious purpose or as an intriguing puzzle, to find two ordinary numbers whose sum is 10 and whose product is 22. In the language of algebra we can choose two letters to stand for the numbers, let us say p and q, and write down:

$$p + q = 10, \qquad pq = 22.$$

Now is our opportunity magically to transform these statements which do **not** tell us directly what the numbers are, or even whether two such numbers exist, into different statements that will answer those equations. What shall we do first? That is the difficulty!

Just as there are no rules to tell us how to look at a geometrical problem, so there are no rules to tell us which moves to make in the game of algebra. Like chess players we must rely on a mixture of judgement, based on past experience, and looking ahead to decide on the best move. Unlike chess players, however, we are not playing with clocks so we cannot get into time trouble and, an even greater advantage, we can take back as many moves as we please, when we please, and start again.

One idea is to turn these two statements about two numbers into one statement about one number, in the hope that it will tell us directly what that number is. This can be done by the following sequence of moves.

$$p + q = 10 \qquad pq = 22$$
$$p^2 + pq = 10p \qquad pq = 22$$
$$p^2 + 22 = 10p$$

Notice that the middle statement is not by itself very enlightening. It is typical of the game of algebra that most positions are not illuminating, but merely exist as stepping stones to the important positions.

Unfortunately for our hopes, the third statement here is a 'quadratic' equation, familiar to generations of school pupils, which can be solved but certainly does not tell us what p might be directly. So let us return to our original position and play something else. The following moves are much superior. If mathematical proofs were annotated like games of chess then the first move would deserve an '!' for its excellence and the subtle idea behind it.

$$p + q = 10 \qquad\qquad pq = 22$$
$$(p + q)^2 = 100 \qquad\qquad pq = 22$$
$$p^2 + 2pq + q^2 = 100 \qquad\qquad 4pq = 88$$
$$p^2 - 2pq + q^2 = 100 - 88 = 12$$
$$(p - q)^2 = 12$$
$$p - q = \sqrt{12}$$

The purpose of this sequence was to turn $(p+q)^2$ into $(p-q)^2$ and hence discover the value of $p-q$. We now know the sum and the difference of p and q and we can easily calculate each of them separately, in three more moves each.

To find p: $\quad p + q = 10$ $\qquad\qquad$ To find q: $\quad p + q = 10$
$\qquad\qquad\quad\ p - q = \sqrt{12}$ $\qquad\qquad\qquad\qquad\quad\ p - q = \sqrt{12}$
$\qquad\qquad\quad\ 2p \ \ = 10 + \sqrt{12}$ $\qquad\qquad\qquad\quad\ 2q = 10 - \sqrt{12}$
$\qquad\qquad\quad\ p \ \ = 5 + \tfrac{1}{2}\sqrt{12}$ $\qquad\qquad\qquad\quad\ q = 5 - \tfrac{1}{2}\sqrt{12}$

This sequence of moves is effective and it is based on an ingenious idea but it is still not the simplest approach, which might be suggested by the form of the solutions, $5 + \sqrt{?}$ and $5 - \sqrt{?}$.

Let us return to the original position for a moment, keep our hands off the pieces, and consider. If two numbers sum to, for example, 10, then one of them exceeds 5 by the amount by which the other falls short of 5. For example, $10 = 6 + 4$, and 6 and 4 are one more than 5 and one less than 5 respectively. With this thought in mind, we will forget about our original choice of p and q, and instead call the two numbers $5 - e$ and $5 + e$. The choice of 'e' is quite arbitrary. Any other letter would do as well.

This is an elegant idea because the sum of the two numbers now automatically comes to 10:

$$5 - e + 5 + e = 10.$$

We therefore only have to concentrate on the fact that their product is 22:

$$(5 - e)(5 + e) = 22.$$

Fortunately this product is very simple and well known. The product of the sum and difference of two numbers is the difference between their squares, so we move immediately to

$$5^2 - e^2 = 22$$
$$25 - e^2 = 22$$

This is much simpler! If $25 - e^2 = 22$, then e^2 must be 3 and e is $\sqrt{3}$. The two numbers are therefore $5 - \sqrt{3}$ and $5 + \sqrt{3}$. Are these the numbers found already? Yes, they are, because $\frac{1}{2}\sqrt{12}$ is equal to $\sqrt{3}$.

Our original sequence of moves took us to a quadratic equation. Does it follow that we can always solve a quadratic by reversing those moves to reach a problem about the sum and product of two numbers? We can indeed, by making one very simple move and then interpreting the result. Starting with,

$$t^2 + 77 = 18t$$

as an example of a quadratic, divide every term by t:

$$t + 77/t = 18.$$

On the left hand side of this last statement are two numbers, t and $77/t$, whose product is 77, while the whole statement says that the sum of these two numbers is 18. So we can find t and $77/t$ by solving the equations

$$x + y = 18, \qquad xy = 77$$

Alternatively, just as before, we can argue that two numbers whose sum is 18 must be $9-c$ and $9+c$ (where c is again an arbitrary choice, which could be exchanged for any other letter except for x or y which would be confusing here). The product of $9-c$ and $9+c$ is 77 and in three more moves we deduce that c is 2:

$$(9 - c)(9 + c) = 77$$
$$81 - c^2 = 77$$
$$c^2 = 4$$
$$c = 2.$$

The two numbers are therefore 2 less and 2 more than 9, or 7 and 11.

These two problems have a feature in common. c^2 here was only 4, and e^2 in the first problem was only 3. What guarantee is there that these squares will not be even less? How can we be certain that solutions to these problems will always exist? We cannot! Suppose that

$$m + n = 12, \qquad mn = 40.$$

Following the same pattern we halve 12 and expect the numbers to be $6-k$ and $6+k$.

$$(6 - k)(6 + k) = 40$$
$$36 - k^2 = 40$$

This is impossible, if k is a real number. This will be no surprise if you recall problem 2f in Chapter 15. If the sum of two numbers is 12 then their product is a maximum when they are both 6, and the product is 36. If the numbers are real, their product can never be 40.

Where is the dividing line to be drawn between problems with solutions, and problems without? The critical comparison is made between the product of the numbers, and the square of half their sum. Therefore if, say, we start from the equations

$$g + h = 14, \qquad gh = 49,$$

in which one half of 14 is 7 and $7^2 = 49$ we ought to be on the boundary itself. We are. This problem has only *one* solution, in which g and h are both equal to 7.

The arguments of this chapter have been entirely algebraic with no appeal to geometrical images of any kind. Yet we clearly see not just the algebraic sentences written on the page, but possibilities and ideas, sequences of moves, good moves and bad moves, not forgetting impossibilities. We can actually 'see' what is impossible, when we are 'seeing' not only literally but metaphorically, with the mind's eye.

The same ambiguity and richness in the meaning of 'see' is observed in everyday life. A recent survey showed that the commonest sense of 'see' was actually 'understand' as in 'I see what you mean!' and I am aware as I write this that 'showed' meaning 'concluded' or 'demonstrated' is another metaphor from the language of sight. Pictures are indeed strictly unnecessary in using the language of algebra but it does not follow that they do not exist or that they cannot be very illuminating, as some of the problems that follow will illustrate.

Problems

1

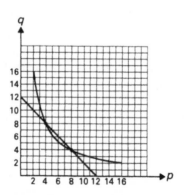

Every pair of numbers p and q, such that $p + q = 12$, corresponds to a point on this straight line, and every pair of numbers such that $pq = 32$ corresponds to a point on the curve. There are two common points where $p = 4$ and $q = 8$, or $p = 8$ and $q = 4$, when both equations are true at the same time.

What geometrical fact about a straight line and this curve corresponds to the fact that this kind of equation can have zero solutions, one solution, or two solutions?

2

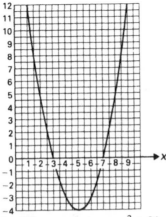

This is a picture of the difference between $x^2 + 21$ and $10x$, so it actually shows the values of $x^2 + 21 - 10x$. When $x^2 + 21$ and $10x$ are equal, their difference will be zero. These two points, corresponding to $x = 3$ and $x = 7$, are marked with small dots.

What geometrical property of the diagram corresponds to the fact that the sum of the two solutions, $3 + 7$, is equal to 10?

3

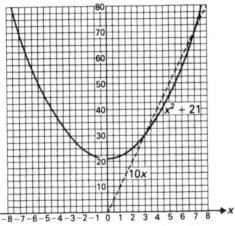

This is a picture of the values of $x^2 + 21$, the curve which is actually a parabola, and $10x$, the straight line. The two points where they are equal, when x is either 3 or 7, are marked with small dots.

Translate the fact that the product of 3 and 7 is 21 into a property of the parabola.

4 When are the two expressions $x^3 + 17x$ and $8x^2 + 10$ equal? What numbers in these expressions are equal to the sum of the answers and the product of the answers?

5 In problem 4, how can 17 be calculated from the numbers which make $x^3 + 17x$ and $8x^2 + 10$ equal to each other?

6 There are no real numbers such that $p + q = 14$ and $pq = 50$. Is it possible to find two numbers which fit these equations by using the complex number 'i', which is equal to $\sqrt{-1}$?

Pattern and illusion

Not all metaphors are instructive. Some of the very worst poetry is packed tight with fancy comparisons and conceits which say nothing true, useful, or effective, but merely exercise the poet's extravagant imagination.

The human brain is adept at spotting resemblances, even when closer inspection reveals that they are not there at all. Recently I watched, horrified, as a cat was struck by a car coming round a bend. I expected to see a corpse lying in the road when the car had passed. I looked for maybe one second. Then my brain reinterpreted the information it had received and I suddenly realised, to my relief, that the 'cat' I had so clearly seen was a piece of dirty paper blown by the wind.

Lewis Carroll wrote the delightful verses which start:

> He thought he saw an Elephant,
> That practised on a fife:
> He looked again, and found it was
> A letter from his wife.
> 'At length I realise,' he said,
> 'The bitterness of Life!'

Adults who smile indulgently at children's riddles get their own daily dose in the punning headlines which are so popular in the newspapers, and which often exploit illusory patterns. The banner headline, 'Getting your kicks from the martial arts' is roughly equivalent to the riddle, 'Why are the martial arts like a stimulant drug?' with the difference that you are given the answer, and expected to appreciate for yourself that the word 'kicks' can also refer to drug use. Whether the psychological thrills of oriental fighting techniques are in any way equivalent to the experience of taking drugs is not obvious. This headline could be a gross libel on martial arts enthusiasts. The epithet, 'The Stern gang' used in a certain well-known newspaper to describe the quartet led by the violinist Isaac Stern is certainly a cheap journalistic trick; the original Stern gang were terrorists, not brilliant musicians.

Misleading or downright false relationships are often spotted by mathematicians. In each of these circles each dot is joined to every other one. Into how many pieces is each circle divided?

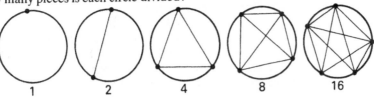

1   2   4   8   16

This problem was invented, or perhaps we should say 'discovered' or 'spotted' by the mathematician Leo Moser. The first circle remains in one part, because the single point cannot be joined to any other. The number of parts then increases: 2, 4, 8, 16, . . .

Surely it is obvious that after five terms, the next term must be 2 × 16 = 32, in order to fit the pattern? It is indeed 'obvious' in the sense that this will be any mathematician's first thought. It is also false.

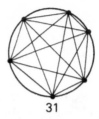

31

The next number is actually 31. After that come 57 and 99. The formula seemed to be: for n points there are 2^{n-1} different regions. The actual formula for the number of regions is

$$\tfrac{1}{24}(n^4 - 6n^3 + 23n^2 - 18n + 24)$$

There is no resemblance at all, apart from the extraordinary match at the very start!

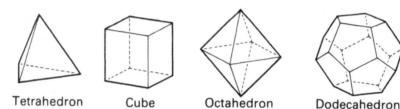

Tetrahedron   Cube   Octahedron   Dodecahedron

Our brains have a natural tendency to spot 'the simplest' explanation for a pattern. Four of the Platonic solids, the regular polyhedra, have 4, 6, 8 and 12 faces respectively. What could be more natural than to guess that a fifth regular solid exists with 10 faces? Or that the sequence continues with 14, 16, . . . faces? This is plausible, but it is also totally false. There is a fifth Platonic solid, the icosahedron, but it has not 10 but 20 faces, and after that there are no others. Is there then no pattern in the numbers 4, 6, 8, 12, 20? Yes, there is, but it is far more complex than 'even numbers in sequence'.

The greatest mathematicians have not been immune to false generalisations. Pierre de Fermat (1601–1665) created the modern theory of numbers and proved dazzling theorems such as: 'A prime number is the sum of the squares of two integers if and only if it is one more than a multiple of 4.' Yet he made a gigantic mistake over what are now called Fermat numbers.

Fermat noticed that

$$2^{2^0} + 1 = 2^1 + 1 = 3$$
$$\text{and } 2^{2^1} + 1 = 2^2 + 1 = 5$$
$$\text{and } 2^{2^2} + 1 = 2^4 + 1 = 17$$
$$\text{and } 2^{2^3} + 1 = 2^8 + 1 = 257$$
$$\text{and } 2^{2^4} + 1 = 2^{16} + 1 = 65\,537$$

are all prime numbers. He expected that $2^{2^5} + 1$ and $2^{2^6} + 1$ . . . and so on, would also be prime. He wrote to Pascal, 'I will answer to you for the truth of this property, but it is not easy to prove, and I confess that I have not yet been able to find a complete demonstration.'

Unfortunately, he was wrong, quite probably completely and totally wrong. Leonard Euler (1707–1783) discovered that $2^{2^5} + 1$ is not prime.

$$2^{2^5} + 1 = 2^{32} + 1 = 4\,294\,967\,297 = 641 \times 6\,700\,417$$

Over a century later, Landry discovered that $2^{2^6} + 1$ is not prime. The numbers are now becoming extraordinarily large, but mathematicians improved their methods of investigation and discovered that the next few 'Fermat numbers' as they were now called, were all composite. With the advent of electronic computers the search has been continued into numbers far greater than the total number of elementary particles in the entire universe, and still no more primes have been found.

The problem is often that too little information is available. How much pattern can be seen in the first sample of this design? A little. In the second sample, much more can be seen, but only the third sample shows the complete pattern, or does it? Can we be certain that the pattern continues for ever, without change? Might it not be just a part of a yet larger pattern?

The Greeks defined a perfect number to be any number equal to the sum of all its factors, including 1 but excluding of course itself. Thus 6 is perfect because $6 = 1 + 2 + 3$ and so is 28, because $28 = 1 + 2 + 4 + 7 + 14$.

The Greeks knew that the first four perfect numbers were 6, 28, 496 and 8128 and it is not surprising that they succumbed to the temptation to see two simple patterns here. It appeared that there was one 1-digit perfect number, one of 2 digits, one of 3 digits and so on, and furthermore that they ended alternately in 6 and 8.

The next perfect number to be discovered was 33550336, recorded anonymously in a medieval manuscript. It destroys the first pattern, but by ending in a 6 it supports the second. Is then the second pattern perhaps correct? No! Thirty perfect numbers are now known and the sequence of their last digits starts 6–8–6–8–6–6–8–8–6–6–8–8–6–8–8. . . It can be proved that the last digit is indeed always either 6 or 8, but that is all. Or is it? Can we exclude the possibility that there is some deeper pattern? No, we cannot, at present. We can only surmise.

All prime numbers are either 1 less than or 1 more than a multiple of 3. In this list of the primes it is very clear that at the beginning at least there are more primes which are 1 less than a multiple of 3:

2 3 5 7 11 13 17 19 23 29 31 37 41 43 47 53 59
61 67 71 73 79 83 89 97 . . .

A check in a table of primes will show that this predominance continues through the hundreds and up into the thousands. This is surely strong evidence that the pattern is genuine and continues for ever! Strong evidence it may seem but not strong enough. There are many patterns among small numbers which break down for larger numbers, and sometimes the 'small' numbers which fit the pattern continue for longer then expected.

The present pattern continues through the hundreds of thousands, through the millions, and tens of millions and hundreds of millions, way up into the billions until the very large number 608981813029 is reached. This is prime and it is 1 **more** than a multiple of 3. At this point the **more** numbers are in the majority for the first time. They continue in the majority until 610968213796 is reached. The next prime is a **less** prime, and so is the one after that. The **less** primes are now in a majority again, but not for ever. It has been proved that the lead changes an infinite number of times in those far reaches of the number sequence which we can only dimly imagine by an exercise of will.

Problems

1 This expression, $4n^3 - 18n^2 + 32n - 15$ is equal to 3, 9, 27, 81 when n has the values 1, 2, 3, 4
  What will its value be when $n = 5$?

2 The fraction $\frac{1}{49}$ expressed as a decimal starts like this :
  0.020 408 163 2. The pairs of digits are the first few powers of 2. Is this pattern genuine, or is it an illusion, a mere coincidence?

3 Is there any pattern in this collection of dots?

4 This arithmetical progression starts with 5 and increases 12 at a time:
  5   17   29   41   53 . . .
  Are all members of this sequence prime numbers? If not, why are there so many to start with?

5 The sum of the angles of a quadrilateral which can be divided into two triangles is 360°.
  When two sides of the quadrilateral cross each other, however, the rule appears to break down. The marked angles in this quadrilateral only sum to approximately 250°. Can the original rule be rescued, so that it applies to every kind of quadrilateral?

6

2	3	5	7	11	13	17	19	23	29	31	37	41	43	47	53
	1	2	2	4	2	4	2	4	6	2	6	4	2	4	6
		1	0	2	2	2	2	2	2	4	4	2	2	2	2	0
			1	2	0	0	0	0	2	0	2	0	0	0	2
				1	2	0	0	0	0	2	2	2	2	0	0	2
					1	2	0	0	0	2	0	0	0	2	0	2	0
						1	2	0	0	2	2	0	0	2	2	2	2	2
							1	2	0	2	0	2	0	2	0	0	0	0
								1	2	2	2	2	2	2	2	0	0	0	2
									1	0	0	0	0	0	0	2	0	0	2

The first row in this table is the sequence of prime numbers. Each row below the first is the sequence of absolute differences between successive numbers in the previous row. 'Absolute' means that all the differences are taken as positive.
  What apparent pattern in this table has attracted the attention of mathematicians?

Clarity, proof and certainty

Why are mathematicians so confident that their mathematics is correct? Why do they sometimes lack confidence? The second question is easier to answer than the first.

Mathematics has much to do with infinity. The counting numbers, after all, go on for ever although the numbers we write down are always relatively small. Galileo puzzled over this pattern.

1	2	3	4	5	6	7	8	9	10	11	12	13	14...
1	4	9	16	25	36	49	64	81	100	121	144	169	196...

The square numbers in the second row are obviously rarer than the counting numbers in the top row. What is more the square numbers obviously become rarer and rarer, as the gaps between them get larger and larger. And yet, Galileo noticed, there is one square number for each counting number. From this point of view the two sequences match perfectly. What is the explanation?

$$1 - \tfrac{1}{2} + \tfrac{1}{3} - \tfrac{1}{4} + \tfrac{1}{5} - \tfrac{1}{6} + \tfrac{1}{7} - \tfrac{1}{8} + \ldots$$
$$1 - \tfrac{1}{2} - \tfrac{1}{4} + \tfrac{1}{3} - \tfrac{1}{6} - \tfrac{1}{8} + \tfrac{1}{5} - \tfrac{1}{10} - \tfrac{1}{12} + \tfrac{1}{7} - \ldots$$

Two hundred years ago series such as these created confusion. At first sight they appear to be almost identical. Indeed, they are almost identical. The same numbers are being added up, with only a small difference in order. Yet this difference turns out to be very significant. The sum of the first series gets closer and closer and closer to the limit 0.693 . . ., while the sum of the second series tends to the limit 0.346 . . ., exactly half as much. A great difference!

Mathematicians solved the problems of summing such series by agreeing to stop talking about sequences of numbers going on for ever 'to infinity' and to talk about finite approximations instead. This does not however get rid of all the problems of infinity. For example, Banach and Tarski proved that it is possible to dissect a small sphere, which might be the size of a pea, into a finite number of parts which can be fitted together to make a much larger sphere, which could be the size of the sun.

Needless to say, their dissection was not into solid lumps each having a definite volume. It is not possible to solve the world's hunger by cutting up one pea and reassembling it to make a mile-wide pea! Their technique is more like Galileo's puzzle in which the square numbers miraculously appear to be as numerous as the counting numbers. The Banach–Tarski paradox would not work if there were not an infinite number of points in a geometric sphere.

In recent years a further source of uncertainty has appeared. It has been exacerbated by the appearance of powerful computers. As more and more mathematicians tackle more and more complex problems, and even use hundreds of hours of computer time to aid their calculation, other mathematicians find it harder to follow what they are doing.

The 4-colour conjecture states, very simply, that a map on a plane surface, or the surface of a sphere, containing only a finite number of regions, can be coloured with 4 colours so that no pair of adjacent regions has the same colour. It was discovered by Francis Guthrie in 1852 and seemed initially rather simple. Several 'proofs' were published, but they all turned out to be fatally flawed.

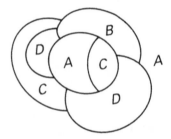

Only recently did Haken and Appel succeed in proving that 4 colours are sufficient by using a computer program to check through 400 different cases. The check was successful and Haken and Appel published their 'proof' but without of course all the details of the program's calculations, which would have filled a book.

Did other mathematicians believe them? Yes, they did, but they could not help noting that the 'proof' was very unusual and set a threatening precedent. Mathematicians do not accept proofs hidden in other mathematicians' heads. How much of a proof could go on inside a computer before other mathematicians cried 'Enough! We are not convinced! Put it down on a few sheets of paper so that we can read it ourselves!'

The most convincing arguments in mathematics neatly sidestep the mysteries of infinity and are the opposite of Haken and Appel's proof. They are short enough for other mathematicians to follow in detail from beginning to end. They take a bird's-eye view of the problem in which all the parts fit together and every individual connection is clear and convincing. The effect is to make the reader say, 'Of course! Now I see it!'

The billiard ball problem starts with a rectangular table divided into a grid of squares, with four corner pockets but no side pockets, and one billiard ball which shoots out of one corner at 45° to the sides and goes bouncing round the cushions until . . . what happens? Will it bounce round the table for ever? Will it end up in one of the pockets? Which pocket? How long will it take to get there? Will its pattern of movement always be the same? Will it strike each cushion the same number of times? Will it . . .?

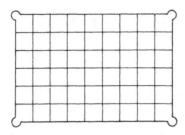

How can sense be made of this variety? To take one specific problem, why and how does the length of the path vary so much? The simplest solution I have seen was shown to me by Shamshad Ersan, a thirteen-year-old schoolboy. It is short, it explains what is happening very clearly, and not surprisingly it involves looking at the problem in a different and special way.

  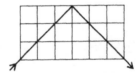

The only facts that are not given are the sides of the rectangle. By experimenting with different choices, always a promising start, we can get some intuitive idea of the possibilities. It does seem that the ball will always end up in a pocket, which is never the one it started from, but the length and pattern of its path varies in a most curious way.

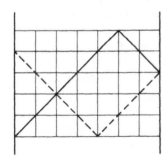 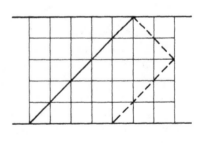

Look at the first 7 by 5 rectangles. Stop watching the ball bouncing from one side to the next side, to the next side . . . and instead watch it moving between the two short ends. Every time it passes from one end to the other it travels 7 units, counting the diagonal of one square as one unit. If it eventually goes into a corner it will have travelled a multiple of 7 units.

Now look at the second rectangle and see the ball bouncing between the top and bottom sides. From one side to the other is 5 units, whether or not it strikes one of the ends along its way. If it finishes in a corner it will have travelled a multiple of 5 units.

Make a list of the distances travelled after 1, 2, 3, 4, . . . lengths of the table and after 1, 2, 3, 4, . . . widths:

0	5	10	15	20	25	30	35
0		7	14	21		28	35

After travelling 35 units, but not before, it will have completed a whole number of lengths and widths and must be in one of the corners. How simple, how convincing and how illuminating!

(Do I hear you asking in which corner it ends up? That is another problem which can also easily be solved by looking in Shamshad Ersan's way.)

Problems

Wilson's theorem bestows a degree of immortality on John Wilson (1741–1793), an English mathematician and lawyer who convinced himself that it was true by testing individual cases but who could not prove it.

It was first published in 1770, and almost immediately proved by the great French mathematician Joseph Louis Lagrange. Ironically, manuscripts in Hannover Library show that Leibnitz had known Wilson's theorem more than one hundred years earlier. In mathematics as in science, inventors who do not publish their discoveries do not get credit for them.

Its statement is very simple. Take any prime number, say 11. Multiply all the integers less than 11 together and add 1:

$$10 \times 9 \times 8 \times 7 \times 6 \times 5 \times 4 \times 3 \times 2 \times 1 + 1 = 3\,628\,801$$

Wilson's theorem states that this total is divisible by 11, the original prime. In fact $3\,628\,801 = 11 \times 329\,891$

Lagrange proved that the converse is also true. If the answer is divisible by the original number, then the original number must be prime.

The following sequence of problems leads step by step to Wilson's theorem. Several of them assume the solution to the previous problem, so each solution should be read before the subsequent problem. If the sequence is successful, then by the end of problem 6 Wilson's theorem will appear surprisingly natural and unexpectedly obvious!

1 Construct a multiplication table for the numbers 0 to 5 and then copy the table, replacing every entry by the remainder when the entry is

divided by 6. So, for example $5 \times 5 = 25$ would be replaced by 1.

Do the same for the numbers 0 to 6, replacing every entry by the remainder when the entry is divided by 7.

What is the connection between the pattern in the tables, and the kind of numbers which 6 and 7 are?

2   The first theorem of higher arithmetic says that if a prime number, p, does not divide m or n, then it does not divide their product $m \times n$. This theorem may possibly seem obvious. In fact it can be proved but we shall assume that it is true.

6	9	1	4	7	10	1	4	7
8	1	5	9	0	4	8	12	5

So, assuming this theorem, why must there be a mistake in these two partial lines which are supposed to be from the multiplication table modulo 11?

3   Why does it follow from problem 2 that each of the numbers less than 11 appears exactly once in each row and each column of the multiplication table modulo 11?

4   In each table, the number 1 appears twice on the long diagonal, at the top end, where $1 \times 1 = 1$, and at the bottom end where $5 \times 5 = 1$ modulo 6, $6 \times 6 = 1$ modulo 7, and $10 \times 10 = 1$ modulo 11.

Why is it not possible for a 1 to appear anywhere else on the long diagonal?

5   If p is a prime, how can all the numbers from 2 to $p - 2$ inclusive be matched in pairs, so that the product of each pair is 1?

6   If p is a prime, what is the product of all the numbers 1 to $p-1$, modulo p? How does Wilson's theorem follow?

Hints for solutions to problems

Introduction

1 Set the original square in a larger square.
2 Suppose that the cube is extremely large.
3 How could you move from the bottom left-hand corner of the base to the top right-hand corner of the base in two moves?
4 Choose a round number between 153 and 482.
5 All of them. How?
6 Try it and see.

Chapter 1

1 A pentagon splits into 3 triangles.
2 A polygon with n sides will make $n - 2$ triangles.
3 The last two. How?
4 Each hexagon is the outline of a cube.
5 The large triangle can be completely filled with copies of the smaller triangle, leaving no space.
6 Think of the triangle as covered with a fine tessellation of small equilateral triangles.

Chapter 2

1 Overlap the two tessellations at right-angles.
2 Either make a tessellation out of the Greek crosses, or use knights' moves, as in chess.
3 Diagonals.
4 Make a model out of cardboard squares, and experiment.
5 Make a model out of cardboard hexagons and triangles, and experiment.
6 Compare problem 3.

Chapter 3

1 A regular tetrahedron has three pairs of opposite edges.
2 Slice the corners off. Where, exactly?
3 The volume of one of the slices cut off the cube is ⅛ of the whole cube.
4 If two solids are of the same shape but one is twice the size of the other in every direction, then the volume of the larger is 8 times that of the smaller.
5 Make a model using cubes of cheese or raw potato.
6 Make a model and experiment.

Chapter 4

There is only one hint for all six problems: 'Try it and see!' In problem 4 this requires a certain amount of visual judgement and fiddling, or the use of compasses to draw arcs with centres on each point, the arcs centred on one point being double the radius of the arcs centred on the other point.

(It is perhaps only fair to admit that there is a strong connection between the solutions to all six problems.)

Chapter 5

1 Replace every other vertex by a small square whose vertices lie on the edges of the original tessellation.
2 The number of squares you can fit in between the edges of the hexagons as you move them apart depends on how far they are moved.
3 Which combinations of regular hexagons and equilateral triangles can possibly fit round each vertex?
4 Try it and see.
5 Try it and see.
6 The hexagonal tile, if it exists, must contain 6 triangles, one of each kind. The pentagon must contain 3 triangles and the quadrilateral 2.

Chapter 6

1 To start the tessellation, take a pair of identical triangles, and join them edge to edge, without turning one of them over.
2 Follow the hint for problem 1. Take a pair of identical quadrilaterals, and join them edge to edge without turning one of them over.
3 The top edge of Abul Wafa's diagram fits neatly against the bottom edge.
4 Cut out two triangles, and use straws or something equivalent to arrange them in perspective from a point X, but with the triangles **not** in the same plane.
5 No hint.
6 Complete the same pattern as far as possible, by finding differences, and two 6s will appear in what would be a row of 6s, if the pattern were complete.

Chapter 7

1 It is essential to think of three dimensional shapes with holes in them, like teacups with handles or doughnuts.
2 Angles.
3 Points on a line.
4 No hint.
5 Experiment with several different pairs of starting numbers.
6 Experiment with several different triangles.

Chapter 8

1 Use the fact from Chapter 1 that a pair of opposite angles sums to 180°.
2 Experiment with a circle which is the easiest conic section to draw (apart from a pair of straight lines).
3 The three circles which get larger and larger become in the limit the sides of a triangle.
4 Draw it and see.
5 Opposite sides of a rectangle are equal, and so are both diagonals.
6 When three similar triangles degenerate into straight lines, they become lines divided in the same ratio.

Chapter 9

1 Add up this large sum, starting with the left-hand end: $1 + 10 = 11$

$$1 + 2 + 3 + 4 + 5 + 6 + 7 + 8 + 9 + 10$$
$$10 + 9 + 8 + 7 + 6 + 5 + 4 + 3 + 2 + 1$$

2 Try rotating the picture, half a turn.
3 $\frac{1}{4}$ is one half of $\frac{1}{2}$; $\frac{1}{16}$ is one half of $\frac{1}{8}$; and so on.
4 $(2 + 1) = 3$
5 The pictures can be drawn so that it is a very symmetrical square.
6 The pieces in the sequence are all L-shaped.

Chapter 10

1 No hint.
2 The end points and the branch points of a tree must be counted together. They are all the ends of edges and correspond to the vertices of the map.
3 Join each of the two points to the origin of the graph.
4 Write the n^2 for the nth squared number as $\frac{1}{2}(2n^2 + 0)$.
5 Experiment with more bubbles and more edges.
6 No hint.

Chapter 11

No hints for these problems.

Chapter 12

1 It is necessary to give twice as much weight to A as to B.
2 From (6, 1) to (16, 7) is 10-right-and-6-up.
3 (3, 2)
4 Draw a diagram. It may help also to find the 'average' of each pair of points.
5 The 'average' of the two points must be given twice as much weight as the third point.
6 All except (10, 4). Why?

Chapter 13

1 Take one kilogram from each of the original weights, and ask where these two separate kilograms would balance.
2 Separate the 2 kg weight into two 1 kg weights, symmetrically balanced about the 2 kg weights' original position.
3 Any pair of weights will be equivalent to a 2 kg weight at the mid-point of their side.
4 How can the point half way between (3, 2) and (7, 4) be calculated?
5 Place the beam on the knife edge so that it almost balances, but not quite.
6 Redistribute 4 kg of the 9 kg weight.

Chapter 14

No hints for these problems.

Chapter 15

1 Dudeney called his solution his 'button and string' method.
2 No hint.

Chapter 16

1 The straight line formed by all the points such that $p + q = ?$ will always be parallel to the straight line in the diagram.
2 If two numbers sum to 10, then their average is 5.
3 Geometrically speaking, 3 and 7 are the distances of the two points where the parabola and the straight line meet, from the vertical axis of symmetry of the parabola.
4 There are three whole numbers less than 10 which make $x^3 + 17x$ equal to $8x^2 + 10$.
5 The calculation from the three numbers which solve problem 3 must be symmetrical. For example, if the numbers are called p, q and r, then $p^2 + q^2 + r^2$ is a plausible (though incorrect) solution because it is symmetrical, but the unsymmetrical $p^2 + qr$ is out of the question.
6 A product such as $(4 - i)(4 + i)$ is equal to
$4^2 - (i)^2 = 16 - (-1) = 16 + 1 = 17$

Chapter 17

1 The value is not $3 \times 81 = 243$.
2 The pattern is genuine, but not in the simple manner in which it starts.
3 Yes.
4 All prime numbers are 1 more or 1 less than a multiple of 6.
5 Count the angles as positive if the line is turning clockwise and negative if it is turning anti-clockwise.
6 The first number in each row of differences appears always to be a 1. Is there any reason for this?

Chapter 18

1 These are the two multiplication tables, with their entries replaced by the remainders on division by 6 and 7 respectively.

×	0	1	2	3	4	5
0	0	0	0	0	0	0
1	0	1	2	3	4	5
2	0	2	4	0	2	4
3	0	3	0	3	0	3
4	0	4	2	0	4	2
5	0	5	4	3	2	1

×	0	1	2	3	4	5	6
0	0	0	0	0	0	0	0
1	0	1	2	3	4	5	6
2	0	2	4	6	1	3	5
3	0	3	6	2	5	1	4
4	0	4	1	5	2	6	3
5	0	5	3	1	6	4	2
6	0	6	5	4	3	2	1

2 The first line contains the numbers 1, 4 and 7 twice each. Why is it impossible for any line in the multiplication table modulo 11 to have the same number twice?
3 How many numbers are there less than 11, to fill each row?
4 Any number on the long diagonal comes from the product of a number with itself, that is, from a perfect square.
5 What pair of numbers produce the 1 in the fourth row?
6 Compared with problem 5, the extra numbers are 1 and $p - 1$.

Solutions to problems

INTRODUCTION

1

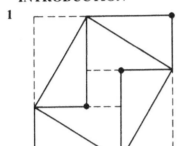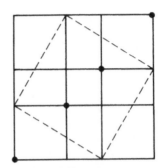

The original figure becomes far more symmetrical when 4 more triangles are added, shown by dotted lines in the first diagram. The symmetry of the complete figure is even clearer in the second diagram, where it is obvious that the four marked points lie on the diagonal of the large square.

2 Six, if the cube is large enough and you are placed inside it, in one corner. (Try standing in the corner of a large rectangular room; you can see all six sides.)

3 The height of the corner furthest above the table is equal to the sum of the heights of the other two corners.

This diagram shows the base of the box, only. *AC* and *BD* are parallel, so in moving from *A* to *C* or from *B* to *D* you will rise by the same amount. Similarly in going either from *A* to *B* or from *C* to *D*. The total of the rises from *A* to *B* and from *A* to *C*, is equal to the rise in moving from *A* to *D*.

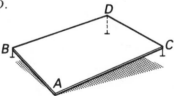

4 Think of a nice round number, say 200, between 153 and 482. From 153 to 200 is 47, and from 200 to 482 is 282. Add 47 and 282 together and the total 'distance' is 329.

(This is by far the easiest method of subtraction to understand, for sums such as this. It is effectively the method used by shopkeepers when they count out your change.)

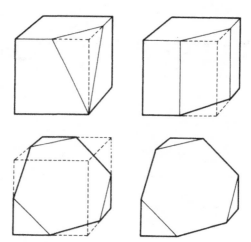

All of them! These figures show how. The triangle can be cut from any sufficiently large cube. There is just one size of cube from which the rectangle could be cut.

The regular hexagon can be cut in four ways, which are equivalent to each other. Its six corners are the middle points of six edges of the cube, and it bisects the cube into two identical halves.

6

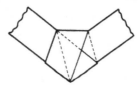

Flatten the knot while gently pulling the ends apart, and this regular pentagon appears.

CHAPTER 1

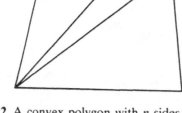

1 Any convex pentagon makes three triangles, which is 6 right-angles or 540°.

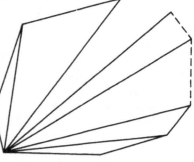

2 A convex polygon with n sides makes $n - 2$ triangles which is $2n - 4$ right-angles or $(180n - 360)°$.

3 The second is the shadow of a cube, looking along a diagonal as in the left-hand figure.

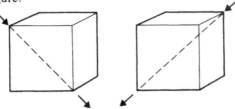

The third is the shadow looking along one of the long diagonals, as in the right-hand figures.

The first shadow is too long in proportion to its width to be the shadow from any direction. It is possible to cut long thin rectangular sections of a cube by slicing off an edge, but these narrow sections do not cast distinct shadows.

4

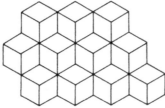

Each hexagon is the outline of one cube. The whole stack looks like this, and produces a well-known optical illusion when your brain cannot decide whether particular cubes are going into or out of the plane of the paper.

5 16 of the smaller triangles fit in like this:

6

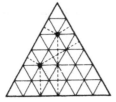

Consider the points in this triangle. They have been chosen to lie on intersections of the triangular grid and therefore the lengths of the lines from each point to the sides are multiples of the height of one of the small triangles.

Moreover, the total length of the lines does not change when moving from one point to the other, because they both lie on a line parallel to one of the sides. However, in at most two moves, it is always possible to move from one intersection on the grid to another by moves parallel to a side. It follows that the sum of the lengths of the lines does not change as long as the points are chosen to lie on intersections of the grid. Since the grid can be chosen as fine as we want, it is possible to come as close as we please to any two points in the triangle, so the conclusion will work whatever two points are chosen.

CHAPTER 2

1

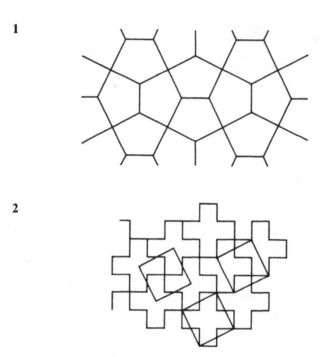

2

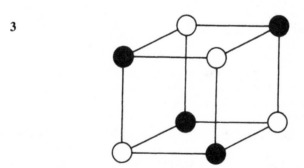

Make a tessellation and join **any** set of corresponding points from four adjacent Greek crosses, as in this diagram.

3

The colours can be switched round, obviously.

4

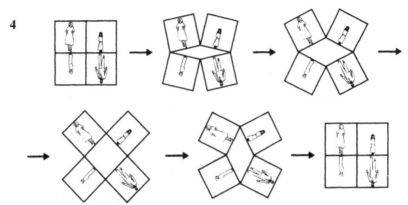

This diagram shows the behaviour of just four adjacent squares. The entire tessellation of squares behaves in the same way.

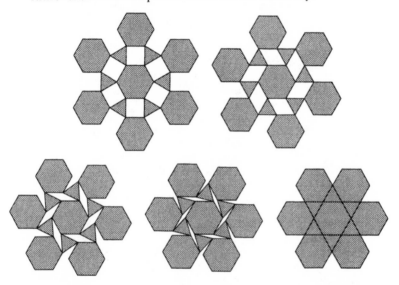

This diagram shows how the hexagons around one hexagon behave. The entire tessellation expands in the same manner, and then closes up again, like the squares in problem 4.

6

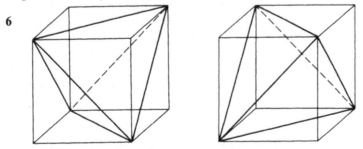

Two tetrahedrons can be inscribed in a cube, depending on which set of diagonally opposite points is chosen for the vertices.

CHAPTER 3

1

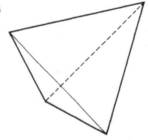

The axes join the mid-points of opposite edges. If a tetrahedron is constructed by slicing corners off a cube as in problem 6 of Chapter 2, then these three axes of the tetrahedron are the lines joining the centres of opposite faces of the cube.

2

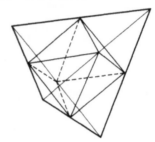

Slice the four corners off, through the middle points of the sides.

3

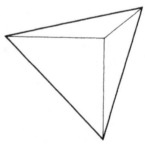

This is the tetrahedron and one of the 4 pyramids sliced off the cube in forming the tetrahedron. Each pyramid has one half of the base of the cube and so is one third of one half, or ⅙ of the whole cube in volume. The four pyramids therefore total ⅔ of the cube and the tetrahedron is the remaining ⅔ or ⅓ of the cube.

4 Slicing the corners off, as in problem 2, produces 5 pieces. Each small tetrahedron is one half of the original tetrahedron in every direction and therefore only ⅛ in volume.

The four small tetrahedra therefore make up exactly half of the original tetrahedron, and the octahedron in the middle makes the other half. Comparing the octahedron with one of the small tetrahedra which have the same edge length, the ratio of their volumes is 4:1.

5

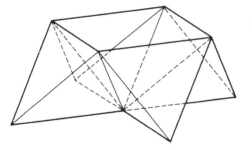

Here are the 4 tetrahedrons inscribed in 4 adjacent cubes. The hollow created is a square pyramid and one half of an octahedron. When the layer above is added a complete octahedron will be formed.

6 Place a regular tetrahedron so that a pair of opposite edges are parallel to a table top and slice it horizontally, half way up. The cross-section is a square.

Since there are three pairs of opposite edges which can be chosen for this operation, there are three possible square cross-sections.

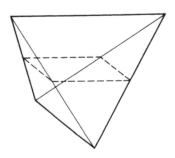
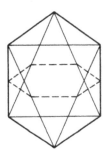

The octahedron can be placed similarly. Rest it on a face and slice it horizontally half way up. Six faces are cut to produce a regular hexagon. There are four ways to place the octahedron and so four hexagonal cross- sections.

CHAPTER 4

1 The corner of the sheet of paper traces out one half of a circle. The points of the pins will be at opposite ends of one diameter.

It is easy to check that it is indeed a circle by placing the point of a compass on the mid-point of the line joining the two drawing pins, which is the centre of the semi-circle.

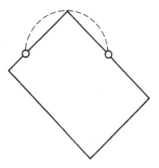

2 The corner traces out a part of a circle. If the angle at the corner is less than a right-angle, it will trace out more than a half-circle, if it is greater than a right-angle it will cover less than half a circle.

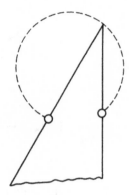

You can confirm that it is indeed a circle by finding the centre quite accurately by trial and error.

3

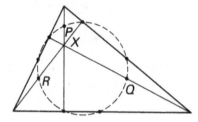

All 9 points lie on a circle. Its centre lies half way between the point X and the centre of the unique circle which goes through the 3 corners of the triangle, and its radius is one half of the radius of that circle.

4

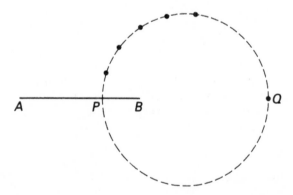

The points all lie on a circle. The two points on the line joining A to B which fit the condition are P and Q in this diagram. The centre of the circle is the mid-point of PQ.

5 Aeneas will enclose the maximum area when his rope forms the shape of a semi-circle against the straight shore line.

6 The area of the quadrilateral will be a maximum when its four vertices lie on a circle.

1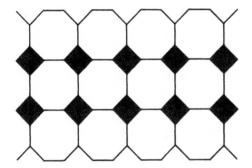

By 'slicing off' just the right amount and replacing each vertex by a square, this tessellation of squares and regular octagons appears.

2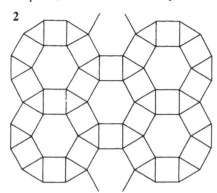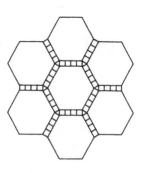

Move the hexagons apart by a distance equal to the length of a side, and one whole square will fit in between each pair of hexagons, with an equilateral triangle filling the space between the vertices.

However, move them apart by a suitably smaller amount, and several squares can be fitted in, in fact any number of squares from two upwards. The diagram shows four squares between each pair of edges. On the other hand by moving them much further apart a row of squares can be fitted in at right-angles to this row.

Only the first solution diagram shows a semi-regular tessellation in which the same regular polygons surround each vertex in the same order.

3

There are just three ways in which regular hexagons and equilateral triangles can fit round a vertex.

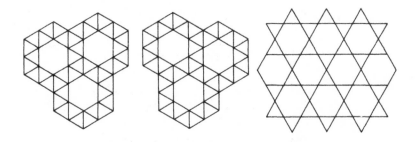

The first possibility does not lead to any semi-regular tessellation. The second leads to two possible tessellations which are mirror images of each other, so many people would say they are essentially the same.

The third possibility leads to a unique tessellation.

4

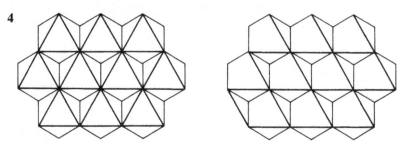

If each selected vertex is joined to the vertices nearest to it, the result is a tessellation of equilateral triangles. However, they might just as well be joined as in the second figure, or in many other ways. There is not a unique way of joining vertices.

5

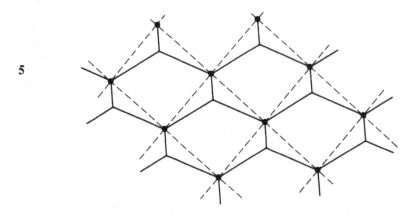

The selected points form a tessellation of squares, if each is joined to the four nearest points, but this way of joining them is not the only possibility.

6 In each case the answer is 'Yes', and in several ways, moreover. Here is one possibility for each part.

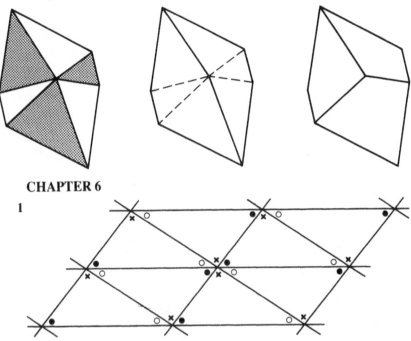

CHAPTER 6

1

Two of each of the three angles of each triangle meet at each vertex, so twice the sum of the angles of each triangle is 360°, and the sum is 180°.

2 One of each of the angles of the quadrilateral appears at each vertex, so their sum is 360°.

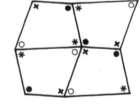

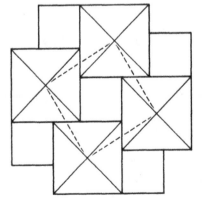

3 Abul Wafa's figure is the tile for another way of looking at the well-known tessellation of two different sizes of squares. The cuts are made by Abul Wafa along the lines joining a set of four corresponding points.

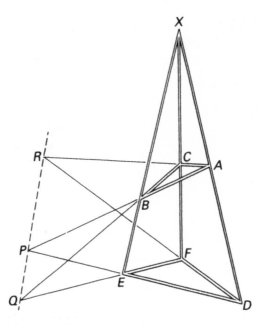

Here is the original figure drawn as a picture of a three-dimensional model. The 3 legs from X form the edges of a triangular pyramid. The top triangle is where the 3 legs from X are sliced by the top plane. The bottom triangle is the base of the pyramid.

The two slicing planes meet along the line L. The edges AB and DE meet at the point P on that line where the top slice, and the plane the pyramid is resting on, and the $ABDE$ side of the pyramid all meet.

The edges BC and EF meet at Q where the top slice and the bottom plane and the $BCEF$ side of the pyramid meet. Similarly the edges CA and FD meet at R.

5 Each row is the differences between the numbers in the row above. To find the next number in the top row, continue the rows of 24s and 0s, and then work backwards by adding 24 to 102, and so on:

0		1		12		63		208		525		1116		2107	
	1		11		51		145		317		591		991		
		10		40		94		172		274		400			
			30		54		78		102		126				
				24		24		24		24		24		24	24	...
					0		0		0		0		0	0	...	

6 Constructing the pattern as far as possible by finding the differences between successive terms, we get this table:

6		15		40		–		162		271		420		615		862
	9		25						109		149		195		247	
		16								40		46		52		
											6		6			
												0				

If the pattern is of the same kind as in problem 5, then the row of zeros and the row of 6s must go right across the table. Working backwards, the missing portion of the table can then be reconstructed, like this:

```
          40     87    162
      25     47     75     109
   16     22    28    34     40
 6   6    6    6    6    6    6    6
   0    0    0    0    0    0    0
```

CHAPTER 7

1 This is a torus, otherwise called a ring or a doughnut. If a circle goes round the torus like this, then the surface of the torus is not divided into two parts and the circle has, as it were, no inside or outside.

2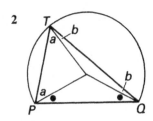

The angle PTQ is always the same. This can be seen if all 3 points are joined to the centre of the circle, making 3 isosceles triangles. Compare problem 2 of Chapter 4.

The angles marked with black dots never change. The sum of the angles of the triangle PTQ never changes, because it is always 180°. So the sum of the angles a, a, b, b never changes. So angle $PTQ = a+b$ never changes.

3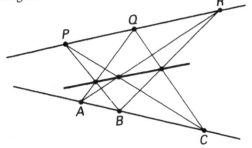

The points where AQ and BP, AR and CP, BR and CQ meet, always lie on a straight line. This is called Pappus' theorem after the Greek geometer who discovered it.

4 The value of a fraction does not change when its numerator and denominator are multiplied or divided by the same number. For example, 84/144 can be reduced to any of the fractions 42/72, 28/48, 21/36, 14/24, 7/12 by dividing numerator and denominator by 2, 3, 4, 6 and 12 respectively.

5 The limiting ratio is invariant with respect to the starting numbers. Whatever starting numbers are chosen the ratio tends to the Golden Ratio.

6 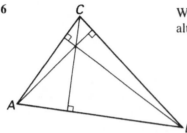 Whatever the initial triangle, the three altitudes always meet in one point.

CHAPTER 8

1 The pair of opposite angles of a cyclic quadrilateral sums to 180 degrees (Chapter 1 pp. 14–15) and since angles *P* and *Q* are equal each must be 90°. Thales, one of the earliest Greek mathematicians, is supposed to have discovered that 'The angle in a semi-circle is a right-angle.'

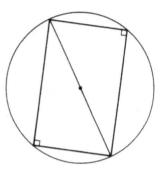

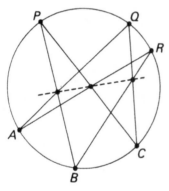

2 Yes, it does work for any conic, though it is only easily tested on circles and ellipses. Here is an illustration.

3

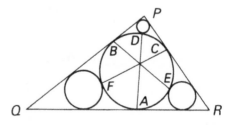

The three circles which get larger and larger tend in the limit to become straight lines, and a figure like this results. *PQR* can be any triangle. As before the lines *AD*, *BE* and *CF* concur.

4

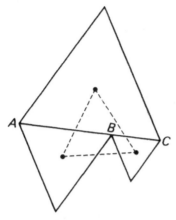

The theorem still works. It just says that if two equilateral triangles are constructed on two parts of the base of another, their centres form a fourth equilateral triangle.

5 Ptolemy's theorem states that $AB \times CD + AD \times BC = AC \times BD$. Since in a rectangle the opposite sides are equal, and so are the diagonals which in this case must be diameters of the circle anyway, Ptolemy's theorem becomes:

$$AB^2 + AD^2 = BD^2$$

This is simply Pythagoras' theorem for the triangle ABD. If we do not know already that BAD is a right-angle because BD is a diameter, then this conclusion would tell us so, because the converse of Pythagoras is also true, that if $AB^2 + AD^2 = BD^2$ then ABD is a right-angle.

6

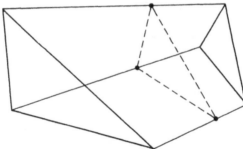

When similar triangles degenerate into straight lines, they become straight lines which are divided in the same ratio. In this figure the corresponding vertices of the original shaded triangles are joined by straight lines which have been divided, by way of illustration, in the ratio 2:1, i.e. two thirds of the way from left to right. As expected, the three points of division are the vertices of another triangle similar to the original shaded pair.

Readers who enjoy drawing their own diagrams may enjoy the experiment of allowing two vertices of the original shaded triangles to move closer together, and then coincide. The result is the generalisation of Napoleon's theorem.

CHAPTER 9

1 Add the series forwards and backwards, like this:

$1 + 2 + 3 + 4 + 5 + 6 + 7 + 8 + 9 + 10$
$10 + 9 + 8 + 7 + 6 + 5 + 4 + 3 + 2 + 1$
$11 + 11 + 11 + 11 + 11 + 11 + 11 + 11 + 11 + 11 = 110$

Therefore the original sum is one half of this, or 55.

2

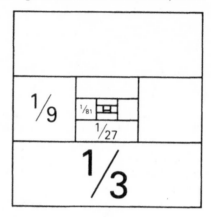

As more and more pieces are filled in, the remaining space is exactly the same shape as the filled area, apart from a small central space, where 'the action' is still taking place.

3

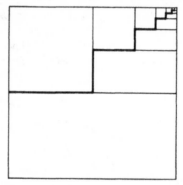

The sequence of squares in the top part matches the sequence of rectangles in the bottom part exactly, except that each square is only one half of the area of its matching rectangle.

So this line divides the whole square into two parts, one of which is double the other. Therefore the top series of squares sums to a total of one third of the square, and the bottom series of rectangles to two thirds.

4 ① + ② + ③ + ④ + ⑤ + ④ + ③ + ② + ①

Take the '1' from the left-hand end, and the '1+2' from the other end, then the '2+3' from the left-hand end, and so on.

The total is $1 + 3 + 5 + 7 + 9$

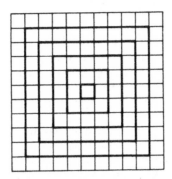

5 The '1' at the start of the series is in the centre. The rings of squares surrounding it have 8, 16, 24, 32 and 40 squares respectively. Five rings, plus the central square make one large square 11 by 11.

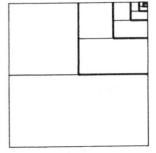

6 This sequence of shapes continues for ever into the top-right corner.
The largest L-shape is ¾ of the whole square. The next largest is 3/16, and so on . . .

CHAPTER 10

1 True. This just states the usual rule for finding the area of a rectangle.

2 True. The end points of a tree and its branch points must be counted together, since they are all ends of edges, in order to correspond with the vertices of the map.

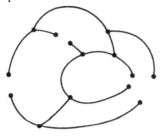

To see the correspondence clearly, cut small chunks out of just enough edges to join all the regions of the map to the outside, so that the map becomes a tree in which the space surrounding the tree counts as one region.

With each chunk that is cut out, one edge becomes two edges, one region disappears, and the number of end points goes up by two. On balance the difference between $V + R$ and $E + 2$ remains zero.

3 Join each of the points to the origin, and then complete a parallelogram, as in this diagram. The far vertex of this parallelogram is the 'sum' of the two original points.

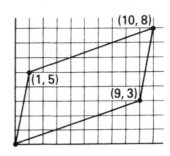

4 By writing n^2 as $\frac{1}{2}(2n^2 + 0)$ the pattern becomes:

triangular	square	pentagonal	hexagonal
$\frac{1}{2}(n^2 + 1n)$	$\frac{1}{2}(2n^2 + 0n)$	$\frac{1}{2}(3n^2 - 1n)$	$\frac{1}{2}(4n^2 - 2n)$

The formula for the hexagonal number simplifies to $2n^2 - n$.

5 The numbers of each are always the same. This can be seen by thinking of the process of placing the bubbles, or the edges of the tree, one at a time. First decide how many bubbles will not be inside any other bubble, and how many edges will meet at the root of the tree.

Thereafter the number of new edges added at the end of an edge matches the number of bubbles placed directly inside a previous bubble.

6 'The average of two numbers is to the two numbers as the position of the number half way between them is to the positions of the two numbers on a ruler or number line.'

(There are obviously other possible wordings.)

CHAPTER 11

1 The argument is only sound as a means of finding an approximate solution. It allows the solver to find the small area which has roughly the same grid-reference on each map. It does not allow exact calculation of the position of the common point, and indeed it does not prove that a common point exists at all.

2

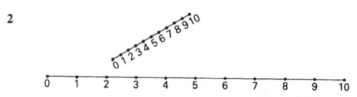

The analogy works, provided the original small map does lie entirely inside the large map, because the argument of the problem still works if the two lines are not parallel, provided a direction is chosen which will define 'matching points'. The one-dimensional argument can then be applied separately to the lines in each map parallel to the top and bottom edges, and to the lines parallel to the left- and right-hand edges.

3 This is true. Sometimes there is no common point if the maps do not overlap sufficiently. To test for the existence of a common point, mark off on the larger map the area corresponding to that part of the smaller map which is 'off the edge'. If the result is two maps, no longer rectangular, such that the smaller is entirely within the larger, then the arguments of the other problems can be applied.

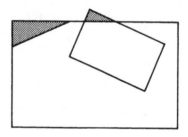

4 This argument is correct, if there is indeed a common point. If sufficiently small maps in the sequence do not overlap the original map at all, this shows that there can be no common point.

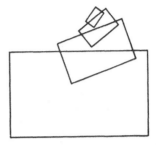

5 This geometrical construction, which has still to be described in detail, will find the common point, if it exists. It says nothing about whether it exists or not.

6 This solution will find the common point if it exists, but it also fails to show whether the point exists or not.

Conclusion: In order to discover whether a common point exists, apply the argument of problem 2 (solution). If the point does exist, then apply any of the last three arguments above to find its exact position.

CHAPTER 12

1 The point sought for is twice as near to A as it is to B, so it is not very surprising that it can be calculated by giving twice as much weight to A as to B. Calculate $2 \times (2, 2) + (8, 14)$ and divide the answer by 3.

$$2 \times (2, 2) + (8, 14) = (12, 18) \text{ and } (12, 18) \div 3 = (4, 6)$$

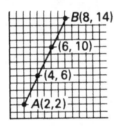

Reversing the roles of A and B will produce the point which is twice as near to B as to A:

$$(2, 2) + 2 \times (8, 14) = (18, 30) \text{ and } (18, 30) \div 3 = (6, 10)$$

2 Yes, they do lie on a straight line, which looks like a staircase with two steps of the same shape, but one is twice the size of the other.

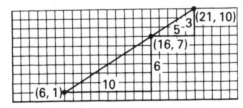

This can be calculated without a drawing like this. From $(6, 1)$ to $(16, 7)$ is 10-right-and-6-up, and from $(16, 7)$ to $(21, 10)$ is 5-right-and-3-up which is exactly half as great a step. So the steps are the same shape and they do lie on a straight line.

153

3 The movements from one point to the next are, 3-right-2-up, 3-right-2-up, 3-right-3-up and 3-right-1-up. The first hiccup is the third movement, 3-right-3-up. If only the point (10, 8) were (10, 7) instead then this would become 3-right-2-up and so would the last movement, to (13, 9). A diagram will check that this gives the answer. (10, 8) is off the line which goes through all the other points.

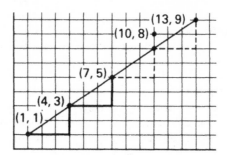

4 If each corner of a triangle is joined by a straight line to the middle of the opposite sides, the three lines all pass through the same point, which is called the centroid of the triangle. It is the point where the triangle would, as it were, balance. The calculation in the question finds the position of this point.

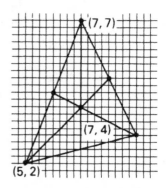

5 Yes, it is, provided that the 'average' of two of the points, (6, 5), is given double the weight of the remaining point (3, 8). So the calculation is:

$$2 \times (6, 5) + (3, 8) = (15, 18) \text{ and } (15, 18) \div 3 = (5, 6)$$

(5, 6) will be the average of the three points, and the centroid of the triangle, whatever pair of points is chosen to have the average (6, 5).

6 Yes, except for the point (10, 4). The movements from (2, 4) to the given points are 7-right-4-up, 4-right-7-up, 1-left-8-up and 8-right-0-up:

By Pythagoras' theorem, the squares of the distances of each point from (2, 4) are $7^2 + 4^2$, $4^2 + 7^2$, $1^2 + 8^2$ and $8^2 + 0^2$. These are all equal to 65 except for the last which is the odd one out.

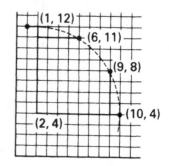

CHAPTER 13

1 Take 1 kg from each. These will be equivalent to 2 kg at the mid-point.

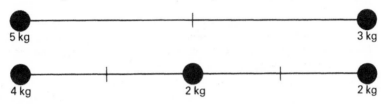

The pair of 2-kg weights are now equivalent to 4 kg as shown.

The pair of 4-kg weights now balance at their mid-point, which is therefore the balance point of the original weights.

2 Separate the 2-kg weight into two 1-kg weights as in this figure, where the line joining the centres has been divided into thirds. The left-hand and right-hand 1-kg weights are now equivalent to a 2-kg weight in the middle, at the same position as the third 1-kg weight.

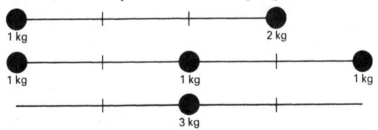

The original weights therefore will balance at a point twice as near to the 2-kg weight as to the 1-kg weight.

3 Any pair of them will be equivalent to a 2-kg weight at the centre of their side.

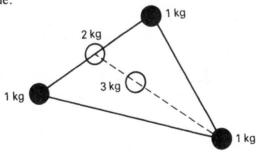

By problem 2, these two weights will balance at a point one third of the way from the 2-kg weight to the other corner.

4 By an actual drawing, the middle point is (5, 3) which can be calculated by taking the numbers half way between 3 and 7 (5), and half way between 2 and 4 (3).

To find the balance point when the third point, (2, 6), is added we can follow problem 2 and find the point one third of the way from (5, 3) to (2, 6). One third of the way from 5 to 2 is 4, and one third of the way from 3 to 6 is also 4. So the points balance at (4, 4).

A more direct but less obvious calculation is to calculate:

$$\tfrac{1}{3}(3+7+2) = 4 \text{ and } \tfrac{1}{3}(2+4+6) = 4$$

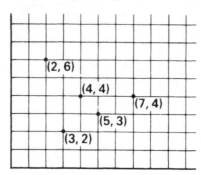

This problem is effectively the same as problem 5 in the previous chapter.

5 Divide the beam into some suitably large number of parts, depending on how heavy it is and how much your small balance will weigh. For example, 11 parts may be enough:

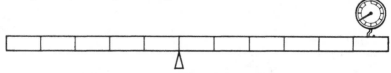

Place the knife edge under the fifth division. Ten of the parts will now balance over the knife edge and all you need to do is to weigh the eleventh part by attaching your balance directly above its middle point. The whole beam will now be in balance, and its total weight is eleven times your reading.

6 Yes, is the simple answer, because the distances of the weights from the knife edge are proportional to the opposite weights. How can this be proved to work?

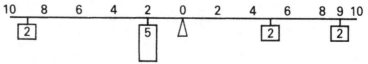

Split off from the 9-kg weight double the opposite weight, 4 kg, and distribute it as in the diagram. The outer pair of 2-kg weights will balance at the knife edge so the problem is reduced to the simpler problem of the remaining 5-kg and 2-kg weights.

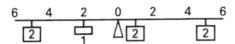

Repeat the process, of splitting off and redistributing a pair of 2-kg weights. The problem is then reduced to a 2-kg weight at 1 and a 1-kg weight at 2, and we know that these weights will indeed balance.

This process applies to any initial pair of weights, whose distances from the knife edge are proportional to the other weight. Each problem is reduced to the same problem with smaller weights . . . and eventually hinges on the very simplest kind of problems which have already been solved.

CHAPTER 14

1 All of them.

2 (a) and (b) are correct but (c) is false since 'difference' could mean the positive differences between them, or the result of subtracting one from the other, in which case the order matters.

3 All of them.

4 Both.

5 (c) Since the description does not state that it is the interior angles which are right-angles there are an infinite number of polygons satisfying it which are not squares. The simplest of these is the Greek cross, composed of 5 identical squares. Here is a more complicated example.

6 All of them, provided that only some properties of negative numbers are to be interpreted. It is not obvious what the product of two negative numbers is to be if each is a temperature, and in none of these cases is it easy to interpret the square root of a negative number.

CHAPTER 15

1 Dudeney used what he called his 'button and string' method. He first drew all the possible knights' moves on the board. These lines are the string.

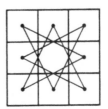

He then pulled the string out to make a simple loop, on which he marked the positions of the four knights.

It is now very easy to see that the minimum solution takes only 7 moves (remembering that successive jumps by the same piece count as one move).

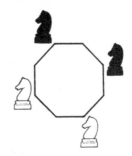

2 (a) and (e) are equivalent, because the eight binary numbers can be placed at the corners of a cube so that adjacent corners have just two pairs of digits in common, like this:

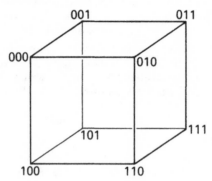

(b) and (d) are equivalent, because it is a property of any quadratic equation when written in the form of this problem that the sum of the two roots is the coefficient on the right-hand side, and their product is the constant term on the left.

For example, in the equation, $p^2 + 12 = 7p$, the two roots sum to 7 and their product is 12, so it is easy to see that they are 3 and 4.
The solution of this type of equation is discussed in Chapter 16.

(c) and (f) are equivalent, because the product of two numbers can be represented as the area of a rectangle whose sides are the two numbers. To say that their sum is 10 is the same as saying that the distance round the rectangle, counting each of its different sides twice, is $2 \times 10 = 20$. The product of the two numbers is a maximum when they are equal, and the rectangle is a square.

CHAPTER 16

1

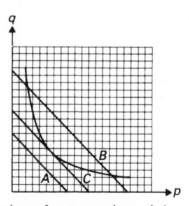

Whatever the values of $p + q$ and pq, their graphs will always be symmetrical about the axis of symmetry in this figure. However, the line will not necessarily cross the curve. Sometimes it will be too far to the lower left (line A), sometimes it will cut the curve in two points (line B) and sometimes it will just touch the curve (line C) in which case there will be only one solution.

2

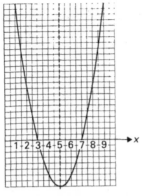

Since the sum of 3 and 7 is 10, the average of 3 and 7 will be one half of 10, or 5. In the figure, $x=5$ marks the axis of the parabola which is indeed half way between the two points where $x^2 + 21 - 10x$ is zero.

3 The distance from the origin to the lowest point of the parabola is 21, wherever the straight line is drawn. In the present case, 3 and 7 are the distances of the two dots from the vertical axis of the parabola, and the geometrical interpretation is that $AB \times CD = 21$.

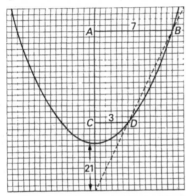

This becomes more interesting if it is realised that the figure of 21 depends only on the parabola. The position of the line makes no difference at all, any line will do.

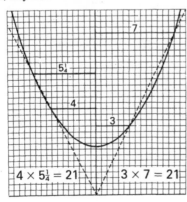

In general this means that if any parabola is drawn and several lines are drawn through the same point on its axis, then the products of the pairs of distances of the intersections from the axis are always the same.

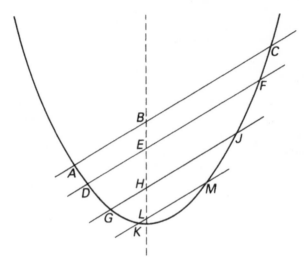

$$BC - AB = EF - DE = HJ - GH = LM - KL$$

Similarly, the property that the sum of the solutions depends only on the line translates into this property of a parabola:

If several lines are drawn crossing a parabola, all parallel to each other, then the sum of the distances of the intersections from the axis is constant. It does not matter whether all distances are measured directly to the axis or all measured along the line. However, distances on different sides of the axis must be counted one positive and the other negative.

4 $x^3 + 17x = 8x^2 + 10$ when x is equal to 1, 2 or 5. The sum of the three solutions is 8 and their product is 10.

5 If 17 can be calculated by a standard rule from 1, 2 and 5, then the calculation must be symmetrical, because each of the solutions 1, 2, 5 is as good as any other. In fact, $17 = 1 \times 2 + 2 \times 5 + 5 \times 1$.

6 Using the idea in the text, solutions if any can be called $7 - x$ and $7 + x$ and their product is 50:

$$(7 - x)(7 + x) = 50$$
$$49 - x^2 = 50$$
$$x^2 = -1$$

and so
$$x = i$$

and the two solutions are $7 - i$ and $7 + i$ (two complex numbers which cannot be marked on the number line).

CHAPTER 17

1 In general, given any four numbers, it is always possible to find a cubic formula, of the same type as $4n^3 - 18n^2 + 32n - 15$ which will produce those four numbers when $n = 1, 2, 3$, and 4.

In this case, the coefficients 4, -18, 32 and -15 were deliberately chosen to produce the values $3, 9, 27, 81$ so that it looks like a sequence of powers of 3, but this is an illusion. When $n = 5$ the expression equals 195.

2 It is true that $\frac{1}{49}$ is equal to the sum of an infinite series involving the powers of 2:

$$\frac{1}{49} = \frac{2}{100} + \frac{4}{10000} + \frac{8}{1000000} + \frac{16}{100000000} + \frac{32}{10000000000} + \frac{64}{1000000000000} + $$
. . .

Consequently the decimal for $\frac{1}{49}$ can be calculated by writing the powers of 2 like this, and adding them up:

$\frac{1}{49} = 0.02$
$\phantom{\frac{1}{49} = 0.}04$
$\phantom{\frac{1}{49} = 0.0}08$
$\phantom{\frac{1}{49} = 0.00}16$
$\phantom{\frac{1}{49} = 0.000}32$
$\phantom{\frac{1}{49} = 0.0000}64$
$\phantom{\frac{1}{49} = 0.00000}128$
$\phantom{\frac{1}{49} = 0.000000}256$
$\phantom{\frac{1}{49} = 0.0000000}512$
$\phantom{\frac{1}{49} = 0.00000000}1024$
$\phantom{\frac{1}{49} = 0.0000000000}. . .$

$0.020408163265306122 . . .$

The pattern in the decimal expression for $\frac{1}{49}$ breaks down, or appears to break down, as soon as the powers of 2, as it were, overlap each other.

3 Yes. The pattern consists of the two copies of a pattern of five dots.

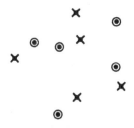

4 No, because the next number in the sequence, 65, is 5×13. Any number which is 2 more or 2 less than a multiple of 6, must be even, and a number 3 more or less than a multiple of 6 must be divisible by 3.

Consequently, all prime numbers are 1 more or 1 less than a multiple of 6. The numbers in the sequence also happen to be each 1 less than a multiple of 6. Since there are quite a lot of small prime numbers, (because there are not many combinations of factors they could have) it is not surprising that $5, 17, 29, 41, 53$ should all be prime.

5 The 'fact' that the angles of a quadrilateral add up to 360° can be partly rescued. One way is to think of an angle as the result of a line turning, and to count clockwise turns as positive, for example, and anti-clockwise turns as negative.

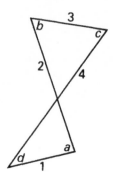

In the diagram, line 1 rotates clockwise to coincide with line 2, so angle a is positive. Line 2 rotates anti-clockwise to coincide with line 3, so angle b is negative. Similarly angle c is negative and angle d is positive.

If we now add the angles up the result will not of course be 360°, but it will in fact be 0°, and a rotation of 0° or 360° means that you end up facing the way you started.

To see that $a+b+c+d = 0$ if b and c are counted as negative, or that $a - b - c + d = 0$ if they are all counted positive, compare the triangles in the next diagram.

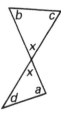

$$b + c + X = 180 = a + X + d \text{ and so } b + c = a + d \text{ or } a - b - c + d = 0$$

6 Mathematicians have spotted, not surprisingly, that the first number in each row of differences is apparently always a 1. This 'fact' has been checked by computer as far as the 63418th row. However, no one has been able to prove that it is always so.

CHAPTER 18

1 A vital property of these multiplication tables which you can check by experimenting with different numbers, is that it makes no difference whether you multiply some numbers together and finally calculate their remainders when divided by 6, or whether you find all their individual remainders and multiply them together and then find the remainder when you divide by 6, or whether you multiply their remainders by using the table.

For example, $7 \times 5 \times 10 = 350$ which leaves remainder 2 when divided by 6. The numbers 7, 5 and 10 have remainders 1, 5 and 4 when divided by 6, and $1 \times 5 \times 4 = 20$ which again leaves a remainder 2 when divided by 6. Finally, from the table, $1 \times 5 = 5$ and $5 \times 4 = 2$. The same answer!

What is the difference in the patterns of the two tables?

Multiplication modulo 6 is a rather lively table, with several isolated zeros and some numbers repeated in some rows and columns.

×	0	1	2	3	4	5
0	0	0	0	0	0	0
1	0	1	2	3	4	5
2	0	2	4	0	2	4
3	0	3	0	3	0	3
4	0	4	2	0	4	2
5	0	5	4	3	2	1

This is the multiplication table modulo 7, which appears less lively than the table modulo 6.

×	0	1	2	3	4	5	6
0	0	0	0	0	0	0	0
1	0	1	2	3	4	5	6
2	0	2	4	6	1	3	5
3	0	0	6	2	5	1	4
4	0	4	1	5	2	6	3
5	0	5	3	1	6	4	2
6	0	6	5	4	3	2	1

The important and relevant difference between 6 and 7 is that 6 is composite, equal to 2×3, while 7 is prime and has no factors, apart from itself and 1.

2

×	0	1	2	3	4	5	6	7	8	9	10
0	0	0	0	0	0	0	0	0	0	0	0
1	0	1	2	3	4	5	6	7	8	9	10
2	0	2	4	6	8	10	1	3	5	7	9
3	0	3	6	9	1	4	7	10	2	5	8
4	0	4	8	1	5	9	2	6	10	3	7
5	0	5	10	4	9	3	8	2	7	1	6
6	0	6	1	7	2	8	3	9	4	10	5
7	0	7	3	10	6	2	9	5	1	8	4
8	0	8	5	2	10	7	4	1	9	6	3
9	0	9	7	5	3	1	10	8	6	4	2
10	0	10	9	8	7	6	5	4	3	2	1

This is the correct multiplication table, modulo 11. Like the table for 7, it contains no zeros outside the zero row and column, and no repetitions in any row or column, because 11 is a prime number. The two rows in the problem had a zero and repeated 1, 4, 5, 7 (not to mention a 12), and so had to be wrong.

163

3 In each row and column of the multiplication table modulo 11, excluding the zero row and column, there are 10 places to be filled, from the numbers less than 11, with no zeros and no repetitions. So each of the numbers 1 to 10 must appear exactly once in each row and column.

4 Suppose for the sake of argument that $n \times n$ leaves a remainder 1 when divided by 11, so that a 1 appeared on the long diagonal where row n and column n cross. Then it would also be true that the difference $n^2 - 1$ was exactly divisible by 11.

Now, by a well-known bit of algebra, $n^2 - 1 = (n + 1)(n - 1)$.
So if $n^2 - 1$ were divisible by 11, the prime number 11 would divide one of the numbers $n + 1$ or $n - 1$. This is impossible because $n + 1$ and $n - 1$ are both less than 11, unless $n + 1 = 11$ and $n = 10$, or $n - 1 = 0$ and $n = 1$, and we already know about these two answers.

5 Imagine the multiplication table, modulo p. Every row contains the numbers 1 to $p - 1$ in some order. In particular every row contains a 1.

1 appears at the start of the first row, and at the end of the last row. Otherwise it does not appear on the long diagonal and each 1 corresponds to a pair of different numbers whose product is 1.

Write down all these pairs and the numbers from 2 to $p - 2$ will all have been included.

Here are the pairs from the modulo 11 table;

$$2 \times 6 = 1 \qquad 3 \times 4 = 1 \qquad 5 \times 9 = 1 \qquad 7 \times 8 = 1$$

6 Imagine once again the table for multiplication modulo p. Match the numbers 2 to $p - 2$ as in the last problem, and multiply them all together. This is the result for $p = 11$:

$$2 \times 3 \times 4 \times 5 \times 6 \times 7 \times 8 \times 9$$
$$= (2 \times 6) \times (3 \times 4) \times (5 \times 9) \times (7 \times 8)$$
$$= 1 \times 1 \times 1 \times 1 = 1$$

Now multiply this product by the remaining two numbers, 1 and $p - 1$, or 10 in this case:

$$1 \times (2 \times 3 \times 4 \times 5 \times 6 \times 7 \times 8 \times 9) \times 10 = 1 \times 1 \times 10 = 10$$

In the general case, the product will be $p - 1$. This is the remainder when $1 \times 2 \times 3 \times 4 \times 5 \times \ldots \times (p - 2) \times (p - 1)$ is divided by p.

What will happen to the remainder if the original number is increased by 1? The remainder will increase by 1 also, so that the remainder will be p.

But wait! If the remainder when we divide by p, is p, then the quotient should actually be 1 greater and then there will be no remainder at all. In other words,

$$1 \times 2 \times 3 \times 4 \times 5 \times 6 \times \ldots \times (p - 2) \times (p - 1) + 1$$

is exactly divisible by p, which is just what Wilson's theorem says.

A CATALOG OF SELECTED
DOVER BOOKS
IN SCIENCE AND MATHEMATICS

Mathematics–Bestsellers

HANDBOOK OF MATHEMATICAL FUNCTIONS: with Formulas, Graphs, and Mathematical Tables, Edited by Milton Abramowitz and Irene A. Stegun. A classic resource for working with special functions, standard trig, and exponential logarithmic definitions and extensions, it features 29 sets of tables, some to as high as 20 places. 1046pp. 8 x 10 1/2. 0-486-61272-4

ABSTRACT AND CONCRETE CATEGORIES: The Joy of Cats, Jiri Adamek, Horst Herrlich, and George E. Strecker. This up-to-date introductory treatment employs category theory to explore the theory of structures. Its unique approach stresses concrete categories and presents a systematic view of factorization structures. Numerous examples. 1990 edition, updated 2004. 528pp. 6 1/8 x 9 1/4. 0-486-46934-4

MATHEMATICS: Its Content, Methods and Meaning, A. D. Aleksandrov, A. N. Kolmogorov, and M. A. Lavrent'ev. Major survey offers comprehensive, coherent discussions of analytic geometry, algebra, differential equations, calculus of variations, functions of a complex variable, prime numbers, linear and non-Euclidean geometry, topology, functional analysis, more. 1963 edition. 1120pp. 5 3/8 x 8 1/2. 0-486-40916-3

INTRODUCTION TO VECTORS AND TENSORS: Second Edition–Two Volumes Bound as One, Ray M. Bowen and C.-C. Wang. Convenient single-volume compilation of two texts offers both introduction and in-depth survey. Geared toward engineering and science students rather than mathematicians, it focuses on physics and engineering applications. 1976 edition. 560pp. 6 1/2 x 9 1/4. 0-486-46914-X

AN INTRODUCTION TO ORTHOGONAL POLYNOMIALS, Theodore S. Chihara. Concise introduction covers general elementary theory, including the representation theorem and distribution functions, continued fractions and chain sequences, the recurrence formula, special functions, and some specific systems. 1978 edition. 272pp. 5 3/8 x 8 1/2. 0-486-47929-3

ADVANCED MATHEMATICS FOR ENGINEERS AND SCIENTISTS, Paul DuChateau. This primary text and supplemental reference focuses on linear algebra, calculus, and ordinary differential equations. Additional topics include partial differential equations and approximation methods. Includes solved problems. 1992 edition. 400pp. 7 1/2 x 9 1/4. 0-486-47930-7

PARTIAL DIFFERENTIAL EQUATIONS FOR SCIENTISTS AND ENGINEERS, Stanley J. Farlow. Practical text shows how to formulate and solve partial differential equations. Coverage of diffusion-type problems, hyperbolic-type problems, elliptic-type problems, numerical and approximate methods. Solution guide available upon request. 1982 edition. 414pp. 6 1/8 x 9 1/4. 0-486-67620-X

VARIATIONAL PRINCIPLES AND FREE-BOUNDARY PROBLEMS, Avner Friedman. Advanced graduate-level text examines variational methods in partial differential equations and illustrates their applications to free-boundary problems. Features detailed statements of standard theory of elliptic and parabolic operators. 1982 edition. 720pp. 6 1/8 x 9 1/4. 0-486-47853-X

LINEAR ANALYSIS AND REPRESENTATION THEORY, Steven A. Gaal. Unified treatment covers topics from the theory of operators and operator algebras on Hilbert spaces; integration and representation theory for topological groups; and the theory of Lie algebras, Lie groups, and transform groups. 1973 edition. 704pp. 6 1/8 x 9 1/4. 0-486-47851-3

Browse over 9,000 books at www.doverpublications.com

A SURVEY OF INDUSTRIAL MATHEMATICS, Charles R. MacCluer. Students learn how to solve problems they'll encounter in their professional lives with this concise single-volume treatment. It employs MATLAB and other strategies to explore typical industrial problems. 2000 edition. 384pp. 5 3/8 x 8 1/2.   0-486-47702-9

NUMBER SYSTEMS AND THE FOUNDATIONS OF ANALYSIS, Elliott Mendelson. Geared toward undergraduate and beginning graduate students, this study explores natural numbers, integers, rational numbers, real numbers, and complex numbers. Numerous exercises and appendixes supplement the text. 1973 edition. 368pp. 5 3/8 x 8 1/2.   0-486-45792-3

A FIRST LOOK AT NUMERICAL FUNCTIONAL ANALYSIS, W. W. Sawyer. Text by renowned educator shows how problems in numerical analysis lead to concepts of functional analysis. Topics include Banach and Hilbert spaces, contraction mappings, convergence, differentiation and integration, and Euclidean space. 1978 edition. 208pp. 5 3/8 x 8 1/2.   0-486-47882-3

FRACTALS, CHAOS, POWER LAWS: Minutes from an Infinite Paradise, Manfred Schroeder. A fascinating exploration of the connections between chaos theory, physics, biology, and mathematics, this book abounds in award-winning computer graphics, optical illusions, and games that clarify memorable insights into self-similarity. 1992 edition. 448pp. 6 1/8 x 9 1/4.   0-486-47204-3

SET THEORY AND THE CONTINUUM PROBLEM, Raymond M. Smullyan and Melvin Fitting. A lucid, elegant, and complete survey of set theory, this three-part treatment explores axiomatic set theory, the consistency of the continuum hypothesis, and forcing and independence results. 1996 edition. 336pp. 6 x 9.   0-486-47484-4

DYNAMICAL SYSTEMS, Shlomo Sternberg. A pioneer in the field of dynamical systems discusses one-dimensional dynamics, differential equations, random walks, iterated function systems, symbolic dynamics, and Markov chains. Supplementary materials include PowerPoint slides and MATLAB exercises. 2010 edition. 272pp. 6 1/8 x 9 1/4.   0-486-47705-3

ORDINARY DIFFERENTIAL EQUATIONS, Morris Tenenbaum and Harry Pollard. Skillfully organized introductory text examines origin of differential equations, then defines basic terms and outlines general solution of a differential equation. Explores integrating factors; dilution and accretion problems; Laplace Transforms; Newton's Interpolation Formulas, more. 818pp. 5 3/8 x 8 1/2.   0-486-64940-7

MATROID THEORY, D. J. A. Welsh. Text by a noted expert describes standard examples and investigation results, using elementary proofs to develop basic matroid properties before advancing to a more sophisticated treatment. Includes numerous exercises. 1976 edition. 448pp. 5 3/8 x 8 1/2.   0-486-47439-9

THE CONCEPT OF A RIEMANN SURFACE, Hermann Weyl. This classic on the general history of functions combines function theory and geometry, forming the basis of the modern approach to analysis, geometry, and topology. 1955 edition. 208pp. 5 3/8 x 8 1/2.   0-486-47004-0

THE LAPLACE TRANSFORM, David Vernon Widder. This volume focuses on the Laplace and Stieltjes transforms, offering a highly theoretical treatment. Topics include fundamental formulas, the moment problem, monotonic functions, and Tauberian theorems. 1941 edition. 416pp. 5 3/8 x 8 1/2.   0-486-47755-X

Browse over 9,000 books at www.doverpublications.com

Mathematics–Logic and Problem Solving

PERPLEXING PUZZLES AND TANTALIZING TEASERS, Martin Gardner. Ninety-three riddles, mazes, illusions, tricky questions, word and picture puzzles, and other challenges offer hours of entertainment for youngsters. Filled with rib-tickling drawings. Solutions. 224pp. 5 3/8 x 8 1/2. 0-486-25637-5

MY BEST MATHEMATICAL AND LOGIC PUZZLES, Martin Gardner. The noted expert selects 70 of his favorite "short" puzzles. Includes The Returning Explorer, The Mutilated Chessboard, Scrambled Box Tops, and dozens more. Complete solutions included. 96pp. 5 3/8 x 8 1/2. 0-486-28152-3

THE LADY OR THE TIGER?: and Other Logic Puzzles, Raymond M. Smullyan. Created by a renowned puzzle master, these whimsically themed challenges involve paradoxes about probability, time, and change; metapuzzles; and self-referentiality. Nineteen chapters advance in difficulty from relatively simple to highly complex. 1982 edition. 240pp. 5 3/8 x 8 1/2. 0-486-47027-X

SATAN, CANTOR AND INFINITY: Mind-Boggling Puzzles, Raymond M. Smullyan. A renowned mathematician tells stories of knights and knaves in an entertaining look at the logical precepts behind infinity, probability, time, and change. Requires a strong background in mathematics. Complete solutions. 288pp. 5 3/8 x 8 1/2.
0-486-47036-9

THE RED BOOK OF MATHEMATICAL PROBLEMS, Kenneth S. Williams and Kenneth Hardy. Handy compilation of 100 practice problems, hints and solutions indispensable for students preparing for the William Lowell Putnam and other mathematical competitions. Preface to the First Edition. Sources. 1988 edition. 192pp. 5 3/8 x 8 1/2. 0-486-69415-1

KING ARTHUR IN SEARCH OF HIS DOG AND OTHER CURIOUS PUZZLES, Raymond M. Smullyan. This fanciful, original collection for readers of all ages features arithmetic puzzles, logic problems related to crime detection, and logic and arithmetic puzzles involving King Arthur and his Dogs of the Round Table. 160pp. 5 3/8 x 8 1/2.
0-486-47435-6

UNDECIDABLE THEORIES: Studies in Logic and the Foundation of Mathematics, Alfred Tarski in collaboration with Andrzej Mostowski and Raphael M. Robinson. This well-known book by the famed logician consists of three treatises: "A General Method in Proofs of Undecidability," "Undecidability and Essential Undecidability in Mathematics," and "Undecidability of the Elementary Theory of Groups." 1953 edition. 112pp. 5 3/8 x 8 1/2. 0-486-47703-7

LOGIC FOR MATHEMATICIANS, J. Barkley Rosser. Examination of essential topics and theorems assumes no background in logic. "Undoubtedly a major addition to the literature of mathematical logic." – *Bulletin of the American Mathematical Society.* 1978 edition. 592pp. 6 1/8 x 9 1/4. 0-486-46898-4

INTRODUCTION TO PROOF IN ABSTRACT MATHEMATICS, Andrew Wohlgemuth. This undergraduate text teaches students what constitutes an acceptable proof, and it develops their ability to do proofs of routine problems as well as those requiring creative insights. 1990 edition. 384pp. 6 1/2 x 9 1/4. 0-486-47854-8

FIRST COURSE IN MATHEMATICAL LOGIC, Patrick Suppes and Shirley Hill. Rigorous introduction is simple enough in presentation and context for wide range of students. Symbolizing sentences; logical inference; truth and validity; truth tables; terms, predicates, universal quantifiers; universal specification and laws of identity; more. 288pp. 5 3/8 x 8 1/2. 0-486-42259-3

Browse over 9,000 books at www.doverpublications.com

Mathematics–Algebra and Calculus

VECTOR CALCULUS, Peter Baxandall and Hans Liebeck. This introductory text offers a rigorous, comprehensive treatment. Classical theorems of vector calculus are amply illustrated with figures, worked examples, physical applications, and exercises with hints and answers. 1986 edition. 560pp. 5 3/8 x 8 1/2. 0-486-46620-5

ADVANCED CALCULUS: An Introduction to Classical Analysis, Louis Brand. A course in analysis that focuses on the functions of a real variable, this text introduces the basic concepts in their simplest setting and illustrates its teachings with numerous examples, theorems, and proofs. 1955 edition. 592pp. 5 3/8 x 8 1/2. 0-486-44548-8

ADVANCED CALCULUS, Avner Friedman. Intended for students who have already completed a one-year course in elementary calculus, this two-part treatment advances from functions of one variable to those of several variables. Solutions. 1971 edition. 432pp. 5 3/8 x 8 1/2. 0-486-45795-8

METHODS OF MATHEMATICS APPLIED TO CALCULUS, PROBABILITY, AND STATISTICS, Richard W. Hamming. This 4-part treatment begins with algebra and analytic geometry and proceeds to an exploration of the calculus of algebraic functions and transcendental functions and applications. 1985 edition. Includes 310 figures and 18 tables. 880pp. 6 1/2 x 9 1/4. 0-486-43945-3

BASIC ALGEBRA I: Second Edition, Nathan Jacobson. A classic text and standard reference for a generation, this volume covers all undergraduate algebra topics, including groups, rings, modules, Galois theory, polynomials, linear algebra, and associative algebra. 1985 edition. 528pp. 6 1/8 x 9 1/4. 0-486-47189-6

BASIC ALGEBRA II: Second Edition, Nathan Jacobson. This classic text and standard reference comprises all subjects of a first-year graduate-level course, including in-depth coverage of groups and polynomials and extensive use of categories and functors. 1989 edition. 704pp. 6 1/8 x 9 1/4. 0-486-47187-X

CALCULUS: An Intuitive and Physical Approach (Second Edition), Morris Kline. Application-oriented introduction relates the subject as closely as possible to science with explorations of the derivative; differentiation and integration of the powers of x; theorems on differentiation, antidifferentiation; the chain rule; trigonometric functions; more. Examples. 1967 edition. 960pp. 6 1/2 x 9 1/4. 0-486-40453-6

ABSTRACT ALGEBRA AND SOLUTION BY RADICALS, John E. Maxfield and Margaret W. Maxfield. Accessible advanced undergraduate-level text starts with groups, rings, fields, and polynomials and advances to Galois theory, radicals and roots of unity, and solution by radicals. Numerous examples, illustrations, exercises, appendixes. 1971 edition. 224pp. 6 1/8 x 9 1/4. 0-486-47723-1

AN INTRODUCTION TO THE THEORY OF LINEAR SPACES, Georgi E. Shilov. Translated by Richard A. Silverman. Introductory treatment offers a clear exposition of algebra, geometry, and analysis as parts of an integrated whole rather than separate subjects. Numerous examples illustrate many different fields, and problems include hints or answers. 1961 edition. 320pp. 5 3/8 x 8 1/2. 0-486-63070-6

LINEAR ALGEBRA, Georgi E. Shilov. Covers determinants, linear spaces, systems of linear equations, linear functions of a vector argument, coordinate transformations, the canonical form of the matrix of a linear operator, bilinear and quadratic forms, and more. 387pp. 5 3/8 x 8 1/2. 0-486-63518-X

Browse over 9,000 books at www.doverpublications.com

Mathematics–Probability and Statistics

BASIC PROBABILITY THEORY, Robert B. Ash. This text emphasizes the probabilistic way of thinking, rather than measure-theoretic concepts. Geared toward advanced undergraduates and graduate students, it features solutions to some of the problems. 1970 edition. 352pp. 5 3/8 x 8 1/2.   0-486-46628-0

PRINCIPLES OF STATISTICS, M. G. Bulmer. Concise description of classical statistics, from basic dice probabilities to modern regression analysis. Equal stress on theory and applications. Moderate difficulty; only basic calculus required. Includes problems with answers. 252pp. 5 5/8 x 8 1/4.   0-486-63760-3

OUTLINE OF BASIC STATISTICS: Dictionary and Formulas, John E. Freund and Frank J. Williams. Handy guide includes a 70-page outline of essential statistical formulas covering grouped and ungrouped data, finite populations, probability, and more, plus over 1,000 clear, concise definitions of statistical terms. 1966 edition. 208pp. 5 3/8 x 8 1/2.   0-486-47769-X

GOOD THINKING: The Foundations of Probability and Its Applications, Irving J. Good. This in-depth treatment of probability theory by a famous British statistician explores Keynesian principles and surveys such topics as Bayesian rationality, corroboration, hypothesis testing, and mathematical tools for induction and simplicity. 1983 edition. 352pp. 5 3/8 x 8 1/2.   0-486-47438-0

INTRODUCTION TO PROBABILITY THEORY WITH CONTEMPORARY APPLICATIONS, Lester L. Helms. Extensive discussions and clear examples, written in plain language, expose students to the rules and methods of probability. Exercises foster problem-solving skills, and all problems feature step-by-step solutions. 1997 edition. 368pp. 6 1/2 x 9 1/4.   0-486-47418-6

CHANCE, LUCK, AND STATISTICS, Horace C. Levinson. In simple, non-technical language, this volume explores the fundamentals governing chance and applies them to sports, government, and business. "Clear and lively ... remarkably accurate." – *Scientific Monthly*. 384pp. 5 3/8 x 8 1/2.   0-486-41997-5

FIFTY CHALLENGING PROBLEMS IN PROBABILITY WITH SOLUTIONS, Frederick Mosteller. Remarkable puzzlers, graded in difficulty, illustrate elementary and advanced aspects of probability. These problems were selected for originality, general interest, or because they demonstrate valuable techniques. Also includes detailed solutions. 88pp. 5 3/8 x 8 1/2.   0-486-65355-2

EXPERIMENTAL STATISTICS, Mary Gibbons Natrella. A handbook for those seeking engineering information and quantitative data for designing, developing, constructing, and testing equipment. Covers the planning of experiments, the analyzing of extreme-value data; and more. 1966 edition. Index. Includes 52 figures and 76 tables. 560pp. 8 3/8 x 11.   0-486-43937-2

STOCHASTIC MODELING: Analysis and Simulation, Barry L. Nelson. Coherent introduction to techniques also offers a guide to the mathematical, numerical, and simulation tools of systems analysis. Includes formulation of models, analysis, and interpretation of results. 1995 edition. 336pp. 6 1/8 x 9 1/4.   0-486-47770-3

INTRODUCTION TO BIOSTATISTICS: Second Edition, Robert R. Sokal and F. James Rohlf. Suitable for undergraduates with a minimal background in mathematics, this introduction ranges from descriptive statistics to fundamental distributions and the testing of hypotheses. Includes numerous worked-out problems and examples. 1987 edition. 384pp. 6 1/8 x 9 1/4.   0-486-46961-1

Mathematics–Geometry and Topology

PROBLEMS AND SOLUTIONS IN EUCLIDEAN GEOMETRY, M. N. Aref and William Wernick. Based on classical principles, this book is intended for a second course in Euclidean geometry and can be used as a refresher. More than 200 problems include hints and solutions. 1968 edition. 272pp. 5 3/8 x 8 1/2. 0-486-47720-7

TOPOLOGY OF 3-MANIFOLDS AND RELATED TOPICS, Edited by M. K. Fort, Jr. With a New Introduction by Daniel Silver. Summaries and full reports from a 1961 conference discuss decompositions and subsets of 3-space; n-manifolds; knot theory; the Poincaré conjecture; and periodic maps and isotopies. Familiarity with algebraic topology required. 1962 edition. 272pp. 6 1/8 x 9 1/4. 0-486-47753-3

POINT SET TOPOLOGY, Steven A. Gaal. Suitable for a complete course in topology, this text also functions as a self-contained treatment for independent study. Additional enrichment materials make it equally valuable as a reference. 1964 edition. 336pp. 5 3/8 x 8 1/2. 0-486-47222-1

INVITATION TO GEOMETRY, Z. A. Melzak. Intended for students of many different backgrounds with only a modest knowledge of mathematics, this text features self-contained chapters that can be adapted to several types of geometry courses. 1983 edition. 240pp. 5 3/8 x 8 1/2. 0-486-46626-4

TOPOLOGY AND GEOMETRY FOR PHYSICISTS, Charles Nash and Siddhartha Sen. Written by physicists for physics students, this text assumes no detailed background in topology or geometry. Topics include differential forms, homotopy, homology, cohomology, fiber bundles, connection and covariant derivatives, and Morse theory. 1983 edition. 320pp. 5 3/8 x 8 1/2. 0-486-47852-1

BEYOND GEOMETRY: Classic Papers from Riemann to Einstein, Edited with an Introduction and Notes by Peter Pesic. This is the only English-language collection of these 8 accessible essays. They trace seminal ideas about the foundations of geometry that led to Einstein's general theory of relativity. 224pp. 6 1/8 x 9 1/4. 0-486-45350-2

GEOMETRY FROM EUCLID TO KNOTS, Saul Stahl. This text provides a historical perspective on plane geometry and covers non-neutral Euclidean geometry, circles and regular polygons, projective geometry, symmetries, inversions, informal topology, and more. Includes 1,000 practice problems. Solutions available. 2003 edition. 480pp. 6 1/8 x 9 1/4. 0-486-47459-3

TOPOLOGICAL VECTOR SPACES, DISTRIBUTIONS AND KERNELS, François Trèves. Extending beyond the boundaries of Hilbert and Banach space theory, this text focuses on key aspects of functional analysis, particularly in regard to solving partial differential equations. 1967 edition. 592pp. 5 3/8 x 8 1/2.
0-486-45352-9

INTRODUCTION TO PROJECTIVE GEOMETRY, C. R. Wylie, Jr. This introductory volume offers strong reinforcement for its teachings, with detailed examples and numerous theorems, proofs, and exercises, plus complete answers to all odd-numbered end-of-chapter problems. 1970 edition. 576pp. 6 1/8 x 9 1/4. 0-486-46895-X

FOUNDATIONS OF GEOMETRY, C. R. Wylie, Jr. Geared toward students preparing to teach high school mathematics, this text explores the principles of Euclidean and non-Euclidean geometry and covers both generalities and specifics of the axiomatic method. 1964 edition. 352pp. 6 x 9. 0-486-47214-0

Mathematics–History

THE WORKS OF ARCHIMEDES, Archimedes. Translated by Sir Thomas Heath. Complete works of ancient geometer feature such topics as the famous problems of the ratio of the areas of a cylinder and an inscribed sphere; the properties of conoids, spheroids, and spirals; more. 326pp. 5 3/8 x 8 1/2. 0-486-42084-1

THE HISTORICAL ROOTS OF ELEMENTARY MATHEMATICS, Lucas N. H. Bunt, Phillip S. Jones, and Jack D. Bedient. Exciting, hands-on approach to understanding fundamental underpinnings of modern arithmetic, algebra, geometry and number systems examines their origins in early Egyptian, Babylonian, and Greek sources. 336pp. 5 3/8 x 8 1/2. 0-486-25563-8

THE THIRTEEN BOOKS OF EUCLID'S ELEMENTS, Euclid. Contains complete English text of all 13 books of the Elements plus critical apparatus analyzing each definition, postulate, and proposition in great detail. Covers textual and linguistic matters; mathematical analyses of Euclid's ideas; classical, medieval, Renaissance and modern commentators; refutations, supports, extrapolations, reinterpretations and historical notes. 995 figures. Total of 1,425pp. All books 5 3/8 x 8 1/2.
Vol. I: 443pp. 0-486-60088-2
Vol. II: 464pp. 0-486-60089-0
Vol. III: 546pp. 0-486-60090-4

A HISTORY OF GREEK MATHEMATICS, Sir Thomas Heath. This authoritative two-volume set that covers the essentials of mathematics and features every landmark innovation and every important figure, including Euclid, Apollonius, and others. 5 3/8 x 8 1/2.
Vol. I: 461pp. 0-486-24073-8
Vol. II: 597pp. 0-486-24074-6

A MANUAL OF GREEK MATHEMATICS, Sir Thomas L. Heath. This concise but thorough history encompasses the enduring contributions of the ancient Greek mathematicians whose works form the basis of most modern mathematics. Discusses Pythagorean arithmetic, Plato, Euclid, more. 1931 edition. 576pp. 5 3/8 x 8 1/2.
0-486-43231-9

CHINESE MATHEMATICS IN THE THIRTEENTH CENTURY, Ulrich Libbrecht. An exploration of the 13th-century mathematician Ch'in, this fascinating book combines what is known of the mathematician's life with a history of his only extant work, the Shu-shu chiu-chang. 1973 edition. 592pp. 5 3/8 x 8 1/2.
0-486-44619-0

PHILOSOPHY OF MATHEMATICS AND DEDUCTIVE STRUCTURE IN EUCLID'S ELEMENTS, Ian Mueller. This text provides an understanding of the classical Greek conception of mathematics as expressed in Euclid's Elements. It focuses on philosophical, foundational, and logical questions and features helpful appendixes. 400pp. 6 1/2 x 9 1/4. 0-486-45300-6

BEYOND GEOMETRY: Classic Papers from Riemann to Einstein, Edited with an Introduction and Notes by Peter Pesic. This is the only English-language collection of these 8 accessible essays. They trace seminal ideas about the foundations of geometry that led to Einstein's general theory of relativity. 224pp. 6 1/8 x 9 1/4. 0-486-45350-2

HISTORY OF MATHEMATICS, David E. Smith. Two-volume history – from Egyptian papyri and medieval maps to modern graphs and diagrams. Non-technical chronological survey with thousands of biographical notes, critical evaluations, and contemporary opinions on over 1,100 mathematicians. 5 3/8 x 8 1/2.
Vol. I: 618pp. 0-486-20429-4
Vol. II: 736pp. 0-486-20430-8

Browse over 9,000 books at www.doverpublications.com

Physics

THEORETICAL NUCLEAR PHYSICS, John M. Blatt and Victor F. Weisskopf. An uncommonly clear and cogent investigation and correlation of key aspects of theoretical nuclear physics by leading experts: the nucleus, nuclear forces, nuclear spectroscopy, two-, three- and four-body problems, nuclear reactions, beta-decay and nuclear shell structure. 896pp. 5 3/8 x 8 1/2. 0-486-66827-4

QUANTUM THEORY, David Bohm. This advanced undergraduate-level text presents the quantum theory in terms of qualitative and imaginative concepts, followed by specific applications worked out in mathematical detail. 655pp. 5 3/8 x 8 1/2.
0-486-65969-0

ATOMIC PHYSICS AND HUMAN KNOWLEDGE, Niels Bohr. Articles and speeches by the Nobel Prize–winning physicist, dating from 1934 to 1958, offer philosophical explorations of the relevance of atomic physics to many areas of human endeavor. 1961 edition. 112pp. 5 3/8 x 8 1/2. 0-486-47928-5

COSMOLOGY, Hermann Bondi. A co-developer of the steady-state theory explores his conception of the expanding universe. This historic book was among the first to present cosmology as a separate branch of physics. 1961 edition. 192pp. 5 3/8 x 8 1/2.
0-486-47483-6

LECTURES ON QUANTUM MECHANICS, Paul A. M. Dirac. Four concise, brilliant lectures on mathematical methods in quantum mechanics from Nobel Prize-winning quantum pioneer build on idea of visualizing quantum theory through the use of classical mechanics. 96pp. 5 3/8 x 8 1/2. 0-486-41713-1

THE PRINCIPLE OF RELATIVITY, Albert Einstein and Frances A. Davis. Eleven papers that forged the general and special theories of relativity include seven papers by Einstein, two by Lorentz, and one each by Minkowski and Weyl. 1923 edition. 240pp. 5 3/8 x 8 1/2. 0-486-60081-5

PHYSICS OF WAVES, William C. Elmore and Mark A. Heald. Ideal as a classroom text or for individual study, this unique one-volume overview of classical wave theory covers wave phenomena of acoustics, optics, electromagnetic radiations, and more. 477pp. 5 3/8 x 8 1/2. 0-486-64926-1

THERMODYNAMICS, Enrico Fermi. In this classic of modern science, the Nobel Laureate presents a clear treatment of systems, the First and Second Laws of Thermodynamics, entropy, thermodynamic potentials, and much more. Calculus required. 160pp. 5 3/8 x 8 1/2. 0-486-60361-X

QUANTUM THEORY OF MANY-PARTICLE SYSTEMS, Alexander L. Fetter and John Dirk Walecka. Self-contained treatment of nonrelativistic many-particle systems discusses both formalism and applications in terms of ground-state (zero-temperature) formalism, finite-temperature formalism, canonical transformations, and applications to physical systems. 1971 edition. 640pp. 5 3/8 x 8 1/2. 0-486-42827-3

QUANTUM MECHANICS AND PATH INTEGRALS: Emended Edition, Richard P. Feynman and Albert R. Hibbs. Emended by Daniel F. Styer. The Nobel Prize–winning physicist presents unique insights into his theory and its applications. Feynman starts with fundamentals and advances to the perturbation method, quantum electrodynamics, and statistical mechanics. 1965 edition, emended in 2005. 384pp. 6 1/8 x 9 1/4. 0-486-47722-3

Browse over 9,000 books at www.doverpublications.com